T0073950

Biomedical Imaging Instrumentation

Biomedical Imaging Instrumentation
Applications in Tissue, Cellular and Molecular Diagnostics

Edited by

Mrutyunjay Suar

KIIT-Technology Business Incubator, Kalinga Institute of Industrial Technology (KIIT-DU), Bhubaneswar, India; School of Biotechnology, Kalinga Institute of Industrial Technology (KIIT-DU), Bhubaneswar, India

Namrata Misra

KIIT-Technology Business Incubator, Kalinga Institute of Industrial Technology (KIIT-DU), Bhubaneswar, India; School of Biotechnology, Kalinga Institute of Industrial Technology (KIIT-DU), Bhubaneswar, India

Neel Sarovar Bhavesh

Transcription Regulation group, International Centre of Genetic Engineering and Biotechnology (ICGEB), New Delhi, India

ACADEMIC PRESS

An imprint of Elsevier

elsevier.com/books-and-journals

ELSEVIER

Academic Press is an imprint of Elsevier
125 London Wall, London EC2Y 5AS, United Kingdom
525 B Street, Suite 1650, San Diego, CA 92101, United States
50 Hampshire Street, 5th Floor, Cambridge, MA 02139, United States
The Boulevard, Langford Lane, Kidlington, Oxford OX5 1GB, United Kingdom

British Library Cataloguing-in-Publication Data
A catalogue record for this book is available from the British Library

Library of Congress Cataloging-in-Publication Data
A catalog record for this book is available from the Library of Congress

ISBN: 978-0-323-85650-8

For Information on all Academic Press publications visit our website at
https://www.elsevier.com/books-and-journals

Publisher: Mara Conner
Acquisitions Editor: Carrie Bolger
Editorial Project Manager: John Leonard
Production Project Manager: Kamesh Ramajogi
Cover Designer: Mark Rogers

Typeset by Aptara, New Delhi, India

Working together
to grow libraries in
developing countries

www.elsevier.com • www.bookaid.org

Contents

CHAPTER 3   Ultrasonography Technology and applications in clinical radiology 33

Syamantak Mookherjee, Devjani Ghosh Shrestha, Skandesh Mohan

CHAPTER 5   Current update about instrumentation and utilization of PET-CT scan in oncology and human diseases.... 67

Rohit Gundamaraju, Chandrabhan Rao, Naresh Poondla

Contributors

Sakir Ahmed
Department of Clinical Immunology & Rheumatology, Kalinga Institute of Medical Sciences, KIIT University, Bhubaneswar.

Prajna Anirvan
Department of Gastroenterology, S.C.B Medical College, Cuttack.

Rini Behera
Department of Conservative Dentistry and Endodontics, Institute of Dental Sciences, Siksha O Anusandhan University, Bhubaneswar, Odisha, India.

Madhusudan Bhat
Department of Pathology, All India Institute of Medical Sciences, New Delhi, India.

Jaydeep Bhattacharya
School of Biotechnology, Jawaharlal Nehru University, New Delhi.

Neel Sarovar Bhavesh
Transcription Regulation group, International Centre of Genetic Engineering and Biotechnology (ICGEB), New Delhi, India.

Venkatesh Chelvam
Department of Biosciences and Biomedical Engineering, Indian Institute of Technology Indore, Khandwa Road, Simrol, Indore, Madhya Pradesh; Department of Chemistry, Indian Institute of Technology Indore, Khandwa Road, Simrol, Indore, Madhya Pradesh.

Amulya Cherukumudi
Department of General Surgery, The Bangalore Hospital, Bangalore; Consultant Surgeon, The Bangalore Hospital, Bangalore, Karnataka.

Manmath Kumar Das
Kalinga Institute of Medical Sciences, KIIT University, Bhubaneswar.

Amit Kumar Dinda
Department of Pathology, All India Institute of Medical Sciences, New Delhi, India.

Harsh A Gandhi
School of Biotechnology, Jawaharlal Nehru University, New Delhi.

Rohit Gundamaraju
ER stress and Mucosal Immunology Team, School of Health Sciences, University of Tasmania, Launceston, Tasmania, Australia.

Subhavna Juneja
School of Biotechnology, Jawaharlal Nehru University, New Delhi.

Sonali Karhana
Department of Pathology, All India Institute of Medical Sciences, New Delhi, India.

Mena Asha Krishnan
Department of Biosciences and Biomedical Engineering, Indian Institute of Technology Indore, Khandwa Road, Simrol, Indore, Madhya Pradesh.

Deepak Kushwaha
Amity Institute of Integrative Science and Health/ Amity Institute of Biotechnology, Amity University, Gurgaon, Haryana.

Gajraj Singh Kushwaha
KIIT-Technology Business Incubator, Kalinga Institute of Industrial Technology (KIIT-DU), Bhubaneswar, India.

Sumeet Suresh Malapure
Nuclear Medicine division, Kasturba Medical College, Manipal Academy of Higher Education, Manipal, Udupi, Karnataka.

Lora Mishra
Department of Conservative Dentistry and Endodontics, Institute of Dental Sciences, Siksha O Anusandhan University, Bhubaneswar, Odisha, India.

Namrata Misra
KIIT-Technology Business Incubator, Kalinga Institute of Industrial Technology (KIIT-DU), Bhubaneswar, India; School of Biotechnology, Kalinga Institute of Industrial Technology (KIIT-DU), Bhubaneswar, India.

Skandesh Mohan
Department of Radiology, Royal Derby and Burton University Hospitals NHS Foundation Trust, Derby, UK.

Syamantak Mookherjee
Department of Radiology, Royal Liverpool and Broadgreen University Hospitals NHS Trust, Liverpool, UK.

Ranjita Ghosh Moulick
Amity Institute of Integrative Science and Health/ Amity Institute of Biotechnology, Amity University, Gurgaon, Haryana.

Anupama Ninawe
Department of Biochemistry, All India Institute of Medical Sciences, New Delhi, India.

Sibi Oommen
Dept. of Nuclear Medicine, Manipal College of Health Professions, Manipal Academy of Higher Education, Manipal, Udupi, Karnataka.

Anita Singh Parihar
Department of Dermatology & Venereology, AIIMS, New Delhi, India.

Satabdi Pattanaik
Department of Conservative Dentistry and Endodontics, Institute of Dental Sciences, Siksha O Anusandhan University, Bhubaneswar, Odisha, India.

Naresh Poondla
Richmond University Medical Center, 355 Bare Avenue, Staten Island, NY, United States of America.

Samudyata C. Prabhuswamimath
Unit for Human Genetics, All India Institute of Speech and Hearing, Manasagangothri, Mysore, India.

Doniparthi Pradeep
Department of Cardio-Thoracic & Vascular Surgery, AIIMS, New Delhi, India.

Chandrabhan Rao
Abhinn Innovation Private Limited, Jaipur, Rajasthan, India.

Naomi Ranjan Singh
Department of Conservative Dentistry and Endodontics, Institute of Dental Sciences, Siksha O Anusandhan University, Bhubaneswar, Odisha, India.

Priyanku Pratik Sharma
Department of Urology, Dispur Hospital, Guwahati.

Devjani Ghosh Shrestha
Department of ENT and Head Neck Surgery, Bhagirathi Neotia Women and Childcare, Centre, Newtown, West Bengal, India.

Mrutyunjay Suar
KIIT-Technology Business Incubator, Kalinga Institute of Industrial Technology (KIIT-DU), Bhubaneswar, India; School of Biotechnology, Kalinga Institute of Industrial Technology (KIIT-DU), Bhubaneswar, India.

Manoj Kumar Tembhre
Department of Cardiac Biochemistry, AIIMS, New Delhi, India.

Joseph Thomas
Department of Plastic Surgery, Kasturba Medical College Manipal, Manipal Academy of Higher Education, Manipal, Karnataka.

Pramod V
Consultant Radiologist, Delta Diagnostic Centre, Bangalore, Karnataka.

Preface

Over several decades biomedical imaging has been undergoing rapid technological advancements from the initial, simple use of X-rays for diagnosis of fractures and detection of foreign bodies, into a plethora of powerful techniques, not only for patient management but also for the study of structure–function relationship of biological processes from the cellular level to the whole-organ level. In simple terms, biomedical imaging involves the sequential process of acquiring, processing, and visualizing structural or functional images of living objects including extraction, and processing of image-related information. Being mostly noninvasive, biomedical imaging also offers precise tracking of metabolites that can be used as biomarkers for disease identification, and for personalized treatment response.

This edited compilation contains a collection of chapters from experts, mostly clinicians and researchers, actively working in the interdisciplinary field of biomedical research. The authors have put together a comprehensive set of both well-established and emerging biomedical imaging technologies used in the laboratory and clinical practice, with a focus on CT scan, ultrasound imaging, magnetic resonance imaging, PET scans, SPECT, mammography, hyperspectral imaging, PA imaging, NIRS, and microscopy imaging. Each chapter presents an overview of the specific imaging technology, along with numerous examples of their application to selected systems. Furthermore, the book discusses advantages and disadvantages of each technology and identifies the technological gaps and research needs for future development. An extensive list of references is provided at the end of each chapter.

Overall, this book presents a broad overview of biomedical imaging tools and research methods. It is useful for graduate students, medical students, researchers, and biomedical engineers to improve their research designs and to help develop innovative projects or new technologies that are truly better, faster, cheaper, noninvasive, and safer for the next generation of users.

Biomedical techniques in cellular and molecular diagnostics: Journey so far and the way forward

Gajraj Singh Kushwaha[a], Neel Sarovar Bhavesh[b], Namrata Misra[a,c], Mrutyunjay Suar[a,c]

[a]KIIT-Technology Business Incubator, Kalinga Institute of Industrial Technology (KIIT-DU), Bhubaneswar, India
[b]Transcription Regulation group, International Centre of Genetic Engineering and Biotechnology (ICGEB), New Delhi, India
[c]School of Biotechnology, Kalinga Institute of Industrial Technology (KIIT-DU), Bhubaneswar, India

1.1 Biomedical imaging technology

Recent development in biomedical imaging technology has improved accuracy and sensitivity for the detection of pathologic conditions. The human body is a very complex system at anatomic and molecular levels, therefore data acquisition and analysis of massive data require computational resources. Hence mathematic algorithms and computation capabilities have further enhanced the image reconstruction process in medical imaging. The innovations of advanced next-generation biomedical image resolution and correlation with the artificial intelligence might lead to improved theragnostic strategy. Altogether, biomedical imaging tools, being noninvasive methods of diagnosis, become an essential part of medical sciences. Biomedical imaging techniques play a very important role in health care, including diagnosis and surgical procedures (Fig. 1.1). These techniques offer a wide range of detection levels ranging from molecule detection to whole-body scanning, thus we can visualize anatomic, structural, and physiologic changes in the body. Most of the instruments used for medical imaging rely on the transmission or reflection principle, which means a source of radiation interacts with substances or tissue of the body and is reflected or transmitted by the respective matter. Subsequently, transmitted or reflected waves are detected by a suitable detector, and an image is generated. During the medical imaging process, major physical phenomena include reflection, transmission, absorption, magnetization, emission, and more, depending on the incidence of energy sources. The major radiation energy sources include x-rays, γ-rays, visible rays, ultraviolet rays, infrared, radio waves, ultrasound waves, among others. Historically, medical imaging techniques were more inclined toward imaging of static state of the body using a radiation source (e.g., x-ray, computed

Biomedical Imaging Instrumentation. DOI: https://doi.org/10.1016/B978-0-323-85650-8.00001-2

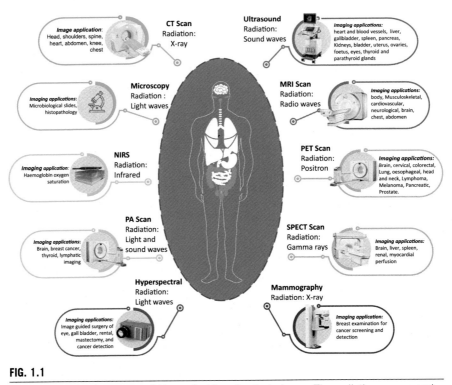

FIG. 1.1

An overview of various types of biomedical imaging techniques. The radiation source and some of their imaging applications are also highlighted.

tomography [CT] scan, and magnetic resonance imaging [MRI]) for musculoskeletal system imaging. However, recent developments in imaging techniques are focusing on the dynamic state that provides more information on metabolic functions (Hendee & Ritenour, 2002). These metabolic functions related to biomedical imaging techniques include nuclear medicine assisted positron emission tomography (PET) and single-photon emission computed tomography (SPECT) scanning, MRI, among others. Besides functional imaging, the integration of machine learning–based artificial intelligence (AI) is getting attention in recent years as it helps in the decision-making process. Moreover, advancement in computational power as well as image processing software is also contributing to image reconstruction. This chapter presents a summary of the major techniques used in medical imaging and biomedical research.

1.1.1 CT scan

CT scan is based on x-ray imaging techniques to produce a depth image of a body. It provides more details compared to a plain x-ray image as imaging process is carried out at several angles to get more information. The CT scan was discovered by Godfrey

Hounsfield and Allan Cormack in 1972, for which they earned the Nobel Prize for this discovery in 1979. The first clinical use of CT scan was carried out in 1974, and now it becomes an essential diagnostic tool for imaging various clinical conditions. The fundamental principle of CT scan is similar to that of traditional x-ray, where x-ray beam is passed through the density of the tissue. However, in CT scanning, it is performed at various angles to obtain the two-dimensional (2D) image. The emitting beams of x-ray is measured from the calculation of the attenuation coefficient, and a reconstruction of the image is developed. In a conventional x-ray, a fixed tube of the x-ray source is used, while CT scanner x-ray source rotates mechanically around a circular opening like gantry. The patient lies on the bed that slowly passes the gantry while the x-ray tube simultaneously moves in a circular path so that x-ray images are generated several times. In comparison with traditional x-ray imaging, CT scan yields 10 times more coarse spatial resolution, at approximately 100 times more than traditional x-ray. The digital x-ray detector is located on the opposite side of the x-ray, and it moves with x-ray sources. Once the rotation is complete, the images are transferred to a computer that reconstructs the 2D image slice of the scan. This process is repeated several times to generate the desired number of image slices. 2D image slices can be stacked to generate a three-dimensional (3D) image to scan an area, which easily differentiates abnormality in bone, muscle, tissue, or fluid. However, sometimes CT images are faint or blurred while imaging soft tissues such as circulatory systems, esophagus, and gastrointestinal (GI) tract. In that situation, contrast agents are intravenously given to patients for enhancement of image contrast. Contrast agents block the passage of x-ray similar to bones, hence more clear images are generated. CT scans are used to acquire detailed cross-sectional images of various internal structures such as bones, soft tissues, internal organs, blood vessels, and the like. Being a very useful imaging technique, CT scan is used in diagnosis and monitoring of various disease conditions such as cancer, heart disease, lung nodules, and liver masses (Ilangovan et al., 2017). Additionally, it is routinely used in the guidance of specific treatment and advanced diagnosis for surgeries, biopsies, and radiation therapy. CT scan utilizes the x-ray sources that produce ionization radiation, hence it may pose a small risk to the fetus in pregnant women. It can also diagnose life-threatening conditions such as hemorrhage, blood clots, and cancer.

1.1.2 Ultrasound imaging

Ultrasound imaging, known as sonography or ultrasonography, is a safe, noninvasive biomedical imaging techniques that generates images by utilizing sound waves. Due to its portability and point-of-care properties, ultrasonography has become an essential diagnostic tool for imaging. Sound is a mechanical wave, and the human ear can hear 20 to 20,000 Hz frequency of sound waves. The sound wave with a frequency of more than 20,000 Hz is called ultrasound waves. Ultrasound waves with frequency 2 to 10 MHz are used in medical imaging ultrasonography. Ultrasound imaging is widely used in gynecology, maternity, abdominal, cardiac, urologic, and tumor cases (Carovac et al., 2011).

The fundamental principle of ultrasound is based on image generation by high-frequency sound waves. The beam of high-frequency sound waves is created by an electric charge passing through a piezoelectric quartz crystal that generates a small potential difference. This phenomenon is known as the piezoelectric effect. The ultrasound beam travels through the body and returns (echo) to the transducer probe that creates a small electric difference in quartz crystals. The potential difference is further amplified by the ultrasound machine that generates ultrasound images on the monitor. The ultrasound waves are generally reflected by air, therefore a transducer is kept in contact with skin. To facilitate the further transmission of sound waves, a gel is applied between the transducer and body skin. The transducer works in both ways (i.e., it produces sound waves and it receives sound waves). When ultrasound waves interact with tissue, two events may occur with waves: attenuation and reflection. Attenuation is the gradual weakening of ultrasound beams while passing through tissue. It may result from various events such as scattering, reflection, or absorption. Reflection happens when ultrasound waves return to the transducer after interacting with the tissue.

Ultrasound imaging information can be represented in various ways, called modes. The following modes are generally used:

A (amplitude) mode: It is the simplest type of display that shows a spiked line generated by echo and plotted as a function of depth on the ultrasound machine monitor. The height of the spike indicates the amplitude of the echo wave.

B (brightness) mode: It is the most commonly used mode of ultrasound in medical imaging and diagnostics. In B mode, ultrasound images are represented as a 2D image generated by dots. The brightness of these dots in the image indicates the amplitude of the returning echo from the body. The complete image is formed by passing multiple ultrasound beams that generate multiple lines of dots.

M (motion) mode: It uses a single line of the beam but in rapid sequence so that it enables measure of a range of motion. M mode is used in echocardiography to study cardiac function and chamber size of the heart.

Doppler mode: It is used to measure blood flow through arteries and veins. It is an important tool in medical imaging for cardiovascular diseases, portal hypertension, blood clots, and more.

Ultrasonography imaging has become one of the most used medical imaging techniques in clinical settings as it images muscle, soft tissues, and even bone surfaces.

1.1.3 Magnetic resonance imaging

MRI is one of the most sophisticated techniques in medical imaging diagnostics that provide a detailed image of body tissue. MRI works on the principle of nucleic magnetic resonance (NMR) that utilizes the magnetic property of hydrogen nuclei, which is part of water and fat present in the body. The hydrogen nucleus, which contains a single proton, has a unique property that acts as a spin, and a positively charged proton creates an electric field. Both these physical properties of hydrogen nuclei result in a tiny magnetic field. Since the body contains many hydrogen nuclei in the form of water, these useful tiny hydrogen magnets are aligned by a strong magnetic field applied by an MR scanning machine. Now, aligned tiny magnets are made to rotate using radio waves, and while returning to equilibrium they release radio frequency (RF) signals, which are used to make images of scanning tissue. MRI is considered a safe medical imaging technique as it does not use radioactive or ionizing radiation. Furthermore, the frequency of radio waves, used in MRI, does not pose any known adverse health effects. The historical progress of MRI imaging is linked with the technical advancements of NMR spectroscopy. The magnetic field was first described by Nikola Tesla in 1882 and NMR was discovered by Felix Bloch and Edward Mills Purcell in 1946 (Bloch et al., 1946; Purcell et al., 1946). Spin echoes and free induction decay were described by Erwin Hahn in 1950 (Hahn, 1950) while one-dimensional (1D) NMR spectrum was reported by Herman Carr in 1952. Later development of pulsed FT-NMR by Richard Ernst revolutionised NMR and now it is ubiquitously used in physical, chemical, biological sciences and medical imaging (Ernst and Anderson, 1966). The imaging process by NMR was first described by Paul C. Lauterbur in 1973 (Lauterbur, 1973). In 2003, Paul Lauterbur and Peter Mansfield were jointly awarded the Nobel Prize in Medicine "for their discoveries concerning magnetic resonance imaging." MRI system comprises four main components that include a magnet, gradient coils, RF coils, amplifiers, receivers, and a computer system (Storey, 2006). Development of Blood-Oxygen-Level Dependent functional MRI (BOLD fMRI) in 1990s (Kwong et al., 1992; Ogawa et al., 1990) allowed noninvasively recording of brain signals with high spatial resolution which led to its applications in Psychology (Berman et al., 2006). Superconducting magnet is the main and biggest part of the MRI scanner that aligns hydrogen nuclei present in the sample. It is a horizontal tubelike structure that creates a strong magnetic field, and the strength of a magnetic field is generally measured in a tesla (T) unit. Commercially available MRI imaging machines are generally equipped with magnets that have magnetic strength between 0.2 and 7 T. Three gradient coils are located within the main magnet, and these coils help to focus on the specific body part for scanning as it decreases or increases the magnetic field on a particular axis (Mualla et al., 2018).

1.1.4 Positron emission tomography scans

PET is a biomedical technique that uses radioactive substances to monitor physiologic activities in the body. PET scanning allows visualizing physiologic changes at the biochemical level in the body utilizing nuclear medicines. The availability of various radioisotope tracer chemicals has increased the application of PET in medical imaging (Berger, 2003). It provides a detailed view of metabolic changes, therefore it is commonly used for imaging of tumors in oncology as well as in neurologic investigations. The working principle of PET scan is that a radiopharmaceutic substance is injected into the patient, and tracer substance is concentrated in the corresponding tissue or organ of interest. Radioactive tracer release positron (β^+) and this subatomic emission are detected by a device called gamma camera. Radioisotopes, used in PET, have short half-lives and undergo decay by emitting a positron. Released positron loses kinetic energy while traveling, and during this process it may interact with an electron. Interaction of these subatomic particles resulted in a high-energy 511 keV annihilation photon that is easily detected by a gamma camera. The most commonly used radiotracer is ^{18}F-labeled fluoro-2-deoxyglucose (^{18}F-FDG). ^{18}F-FDG is a radioisotope analogue of naturally occurring glucose, therefore it is metabolized in different parts of the body. Moreover, the half-life of ^{18}F-FDG is 110 minutes, and a typical PET scanning procedure is completed in 10 to 40 minutes. Apart from ^{18}F-FDG, there are other radionuclides used in PET scanning, including ^{11}C, ^{13}N, and ^{31}Ga, as they have short half-lives. PET isotopes are generally labeled that are metabolized by a human body such as glucose, water, or drugs that bind to specific receptors. PET scanning is mostly used in oncology and neurology investigations. Since it is very useful for quantitative physiology, its application in preclinical studies is growing.

1.1.5 Single-photon emission computerized tomography

SPECT is a biomedical imaging technique similar to PET scanning as both utilize radioisotope tracer to detect signals. However, the major difference is the use of trace substances; SPECT utilizes ^{123}I or ^{131}I, while PET scanning is generally based on ^{18}F (Rahmim & Zaidi, 2008). In SPECT, gamma radiation is released instead of the positron; hence this gamma radiation is detected by a radiation detector. Typically, SPECT detector consists of a scintillation device that converts gamma ray photons to electrical signals. The scintillation camera process signals in three steps: (1) lead collimator for directional modulation of incoming waves; (2) thallium-activated sodium iodide (NaI[Tl]) crystal that converts gamma rays to lower energy photons; and (3) photons converted to electrical signals by a photomultiplier tube (PMT). The desired properties of SPECT instruments should include high intrinsic efficiency, good spatial resolution, and good energy resolution. The technical advancements in a compact design and high-quality detector system make SPECT scanning system more useful. Additionally, image reconstruction algorithm addition provides imaging with better resolution. The 2D image is generated by using these electrical signals. SPECT provides 2D images similar to CT scans while utilizing information on radiopharmaceutic metabolism. Recently, SPECT combination with CT scan gives more useful images, particularly anatomic imaging concerning functional properties (Livieratos, 2015). SPECT scanning is used in many clinical areas, including brain, neurologic, cardiology, and musculoskeletal imaging.

1.1.6 Mammography

Mammography is a noninvasive ray imaging technique used for screening and detection of early breast cancer in women. The working principle is very similar to traditional x-ray imaging in which a body part is exposed to x-ray radiation and transmitted rays are detected as image. X-ray radiation is a type of electromagnetic radiation that can pass through the body forming an image on photographic film or a special detector. X-rays differentially penetrate the body parts because different body parts absorb x-rays. Bones absorb a higher amount of radiation than tissue or organ, thus they appear white on x-ray film whereas tissue or organ appear dark grey or white. The image film of the detector is called a mammogram. The first mammogram was published by German surgeon Albert Salomon in 1913. The dedicated system for mammography was developed by Charles Gross in 1965, equipped with a molybdenum x-ray tube. In 1985, Laszlo Tabar and his team carried out the systematic mammographic screening of 134,867 patients and reported the results. In earlier years, x-ray images were recorded on photographic film but now these are recorded electronically, hence the name digital x-ray (Schmidt & Nishikawa 1995). A mammography machine is a rectangular compact machine with an x-ray generator that emits low-energy x-rays. Usually this machine is designed to image the breast only so there is a device connected to it that holds the breast to take images at different angles. Moreover, the breast is compressed during the mammography so that motion-induced blurring images can be minimized. Compression also helps to improve quality image because of x-ray travel through a shorter path. Mammography is a very useful imaging technique for screening and earlier detection of breast cancer and it allows locating accurately size and shape of breast abnormalities. A mammogram is examined by a radiologist who looks for abnormal areas different from normal tissue. Brighter tiny regions have differential contrast, size, shape, edges, or margin of the region for the possibility of malignancy detection.

1.1.7 Hyperspectral imaging

Hyperspectral imaging (HSI) is a spectrometry-based imaging technique that is emerging as an imaging technique in biomedical diagnosis and research. HSI was originally developed for nonmedical application areas such as remote sensing, archaeology, food analysis, and environmental analysis. Its medical application is currently limited to surgical guidance, hence this technique is comparatively underutilized for diagnostic purposes. As its name indicates, the hyperspectral imaging system involves two major technologies (spectrometry and imaging); therefore it is also called imaging spectrometry (Calin et al., 2014). Three major components (a light source, a monochromator or spectral separator, an area detector) create the HSI. The light source is directed onto the body surface and it travels through body tissue resulting in absorption, transmission, or reflection. Reflected light waves are passed through a spectral separator or collimator where the waves are separated into spectral bands. These spectral bands of reflected waves hit an area detector that is generally made of a conventional charge-couple device (CCD); subsequently, pixels of these spectra are converted to a 2D image. The penetration of light into tissue undergoes specific physical phenomena during its travel as the tissue has a different composition of various kinds of subcellular organelles, macromolecules, and biochemicals that exhibit differential scattering of incident rays (Lu & Fei, 2014). HSI is a noninvasive imaging technique, so it has potential applications in clinical diagnosis. Currently it is being used for the surgical guidance and detection of cancer, cardiac diseases, and diabetic foot ulcers. In image-guided surgical procedures it has been explored in abdominal, renal, and gall bladder surgery as well as mastectomy.

1.1.8 Photoacoustic (PA) imaging

PA imaging is a relatively new technique for biomedical imaging. As its name suggests, it makes use of laser light waves and sound waves, thus it is also called optoacoustic imaging. PA effect was first discovered by Alexander Graham Bell in 1880 by his observation on the production of sound waves by sunlight. However, research work on its biomedical imaging application emerged much later, in the mid-2000s, because of the development in light source technologies. Moreover, the image reconstruction algorithm added more advantages for real-time, high-resolution, and precise reconstruction of images. Laser light is delivered through tissue, and some of these waves are reflected as heat energy leading to transient thermoelastic ultrasonic emission that is measured by ultrasonic transducers. The sensitivity of PA scanning relies on high spatial/temporal resolution with depth imaging that can be further enhanced by using various types of chromophores. PA imaging chromophores can be either exogenous or endogenous. Endogenous chromophores are naturally present in the body such as water, hemoglobin, melanin, and lipids, which are responsive to optic waves. Common exogenous chromophores include small molecule dyes such as methylene blue, indocyanine green, and nanoparticle-based compounds. These chromophores provide a great spectral resolution to PA imaging that makes it a good clinically important noninvasive technique (Beard, 2011). The important clinical applications of PA scans are limited to image superficial tissue because the penetration power of PA is still a major challenge (Steinberg et al., 2019). Hence PA is used in scanning of dermatologic, breast, thyroid, brain, gynecologic, and other organs for malignancy pathology. However, there are still challenges associated with PA imaging.

1.1.9 Near-infrared spectroscopy (NIRS)

NIRS is an infrared radiation spectroscopy-based analytic technique that is being used for biomedical imaging in clinical settings. NIR uses radiation waves that correspond to the near-infrared region of the electromagnetic spectrum (from 780–2500 nm). The principle of NIR is relying on molecular vibrational photons that employ photon energy ranging from 2.65×10^{-19} to 7.96×10^{-20} J. Atoms, present in the molecules, are connected through chemical bond release vibrational energy due to the motion of atoms. Two types of spectral bands, overtones and combination bands, are found in the midregion of NIR electromagnetic spectrum. However, these bands are generally found as an overlapping region in NIR spectrum, therefore various mathematical and statistical methods are employed to detect a meaningful signal. These statistical tools include multivariate calibration, principal component analysis, and artificial neural networks. These tools for calibration of broadband spectra in NIR are called chemometrics. The absorption of NIR spectra is related to the chemical composition of a sample that is being investigated because different chemical bonds may exist in different chemicals. These bonds have variable energy required for the vibration of bonds, thus characteristic bonds can serve as a fingerprint for that compound. The absorption energy for vibration for a specific bond provides a distinct spectrum, and by utilizing chemometric tools, structural properties of the sample are identified. Biomedical application of NIR spectroscopy is limited, and currently it is being used to detect oxygen saturation level of hemoglobin in clinical settings (Sakudo, 2016). However, in vitro applications of NIR spectroscopy in biologic sample analysis are emerging slowly.

1.1.10 Microscopy imaging

Microscopy imaging is an area of biomedical science that deals with the application of a microscope to visualize objects that cannot be seen by the naked eye. There are two radiation source technologies (optical and electron) used in microscopy. Microscopy has revolutionized biomedical imaging, particularly microbiology and histopathology investigations in disease diagnosis. The invention of the microscope began in the 17th century, as Hans and Zacharias Jensen placed one lens behind the other. The first observation of biologic material was by Antonie van Leeuwenhoek. He developed a compound microscope using two lenses that provided a good range of magnification. The developmental progress of microscopes has been associated with the development of suitable materials and methods for fixation, staining dyes, embedding, and preservation. When an object comes closer to the eye, objects become clear with more details. However, there is a limit of this distance where the eye cannot see an object clearly beyond this limit. In general, the adult human eye has a limit of 250 mm or 25 cm to see the object. Additionally, one more property of an eye lens is to identify two points apart in the object. This ability to distinguish two points in an object is called resolution power. Microscopy may be categorized into two classes based on radiation source (optical or light microscope and electron microscope). A light microscope uses visible light as a radiation source, while an electron beam is thrown at the object in an electron microscope. In light microscopy, visible light passes through an object, it is transmitted or reflected through lenses, and subsequently a magnified image of an object is formed. This image can be visualized by the naked eye, photographic film, or digitally on a computer screen using a CCD camera (Coltharp & Xiao, 2012). Light microscopy has further been modified to improve visualization by employing various illumination phenomena. In bright field microscopy, image contrast is generated by the absorbance of light by the sample, while dark field microscopy is based on the scattering of light by a sample. Microscopy is a versatile technology; therefore it is being used in many areas of biomedical sciences, including microbiologic and pathologic applications in clinical settings, academic training, and research.

1.2 Way forward

Currently biomedical engineering research and development is experiencing rapid transformations with the integration of modern image processing algorithms, nanotechnologic advancements, and personalized medicine. The major advantage of biomedical devices over contemporary diagnostic techniques is their nature of noninvasiveness; therefore these techniques are being used routinely in health care settings. The future development in the medical device research area is focusing more on the use of artificial intelligence and machine learning approaches with the integration of the internet of things in the data processing. These information-processing technologies certainly will improve the capability of the device that will subsequently assist in the decision-making process. The second major advancement is focused on wearable medical devices with the integration of the internet of things that provide real-time monitoring of desired health parameters. Altogether, biomedical instruments contribute majorly to the health care sector, particularly for diagnostic applications.

References

Beard, P., 2011. Biomedical photoacoustic imaging. Interface Focus 1, 602–631.

Berger, A., 2003. How does it work? Positron emission tomography. BMJ 326 (7404), 1449.

Berman, M.G., Jonides, J., Nee, D.E., 2006. Studying mind and brain with fMRI. Soc Cogn Affect Neurosci 1749–5024, 1 (2), 158–161. doi:10.1093/scan/nsl019.

Bloch, F., Hansen, W., Packard, M., 1946. Nuclear Induction. Physical Review 70, 460–473. doi:10.1103/PhysRev.70.460.

Calin, M.A., Parasca, S.V., Savastru, D., Manea, D., 2014. Hyperspectral imaging in the medical field: Present and future. Applied Spectroscopy Reviews 49 (6), 435–447.

Carovac, A., Smajlovic, F., Junuzovic, D., 2011. Application of ultrasound in medicine. Acta Informatica Medica 19, 168–171.

Coltharp, C., Xiao, J., 2012. Superresolution microscopy for microbiology. Cell Microbiology 14, 1808–1818.

Ernst, R.R., Anderson, W.A., 1966. Application of Fourier Transform Spectroscopy to Magnetic Resonance. Review of Scientific Instruments 37 (1), 93–102. doi:10.1063/1.1719961.

Hahn, E.L., 1950. Spin Echoes. Physical Reviews 80 (4), 580. doi:10.1103/PhysRev.80.580.

Hendee, W.R., Ritenour, E.R., 2002. Medical imaging physics, 4th ed. John Wiley & Sons.

Kwong, K.K., Belliveau, J.W., Chesler, D.A., Goldberg, I.E., Weisskoff, R.M., Poncelet, B.P., Kennedy, D.N., Hoppel, B.E., Cohen, M.S., Turner, R., 1992. Dynamic magnetic resonance imaging of human brain activity during primary sensory stimulation. Proc Natl Acad Sci U S A 0027-8424, 89 (12), 5675–5679. doi:10.1073/pnas.89.12.5675.

Ilangovan, S.S., Mahanty, B., Sen, S., 2017. Biomedical imaging techniques information Resources Management Association Medical imaging: Concepts, methodologies, tools, and applications. IGI Global, pp. 413–434.

Lauterbur, P.C., 1973. Image Formation by Induced Local Interactions: Examples Employing Nuclear Magnetic Resonance. Nature 242, 190–191. doi:10.1038/242190a0.

Livieratos, L., 2015. Technical pitfalls and limitations of SPECT/CT. Seminars in Nuclear Medicine 45, 530–540.

Lu, G., Fei, B., 2014. Medical hyperspectral imaging: A review. Journal of Biomedical Optics 19, 10901.

Mualla, F., Aubreville, M., Maier, A, et al., 2018. Microscopy. In: Maier, A., Steidl, S., Christlein, V. et al (Eds.). Medical imaging systems: An introductory guide, 5. Springer.

Ogawa, S., Lee, T.M., Nayak, A.S., Glynn, P., 1990. Oxygenation-sensitive contrast in magnetic resonance image of rodent brain at high magnetic fields. Magn Reson Med 0740-3194, 14 (1), 68–78. doi:10.1002/mrm.1910140108.

Purcell, E.M., Torrey, H.C., Pound, R.V., 1946. Resonance Absorption by Nuclear Magnetic Moments in a Solid. Physical Review 69 (1–2), 37–38. doi:10.1103/PhysRev.69.37.

Rahmim, A., Zaidi, H., 2008. PET versus SPECT: Strengths, limitations and challenges. Nuclear Medicine Communications 29, 193–207.

Sakudo, A., 2016. Near-infrared spectroscopy for medical applications: Current status and future perspectives. Clinica Chimica Acta 455, 181–188.

Schmidt, R.A., Nishikawa, R.M., 1995. Clinical use of digital mammography: The present and the prospects. Journal of Digital Imaging 8 (1), S74–S79.

Steinberg, I., Huland, D.M., Vermesh, O., Frostig, H.E., Tummers, W.S., Gambhir, S.S., 2019. Photoacoustic clinical imaging. Photoacoustics 14, 77–98.

Storey, P., 2006. Introduction to magnetic resonance imaging and spectroscopy. Methods in Molecular Medicine 124, 3–57.

Role of CT scan in medical and dental imaging

Lora Mishra, Rini Behera, Satabdi Pattanaik, Naomi Ranjan Singh

Department of Conservative Dentistry and Endodontics, Institute of Dental Sciences, Siksha O Anusandhan University, Bhubaneswar, Odisha, India

2.1 Introduction to computed tomography

The computed tomography (CT) scanner was invented by Godfrey Newbold Hounsfield in Hayes, England. A CT or computerized axial tomography (CAT) amalgamates information from various x-rays to generate a detailed image of internal organs of the human body. Together with direct imaging, CT enables differentiation of soft tissue structures as well. The initial invention of CT dealt with head scans only, but later advancement proved possibilities of whole-body scans. It can be utilized to image almost all anatomic areas, including those susceptible to patient breathing and motion. CT scan has a key role in the reduction of surgeries, and the percentage is markedly reduced from 13% to 5% because CT provides an alternative treatment option for the patient as it describes the patient's condition very precisely. Advancement of techniques in CT such as cone beam (CBCT), dual-energy (DECT), extreme multidetector (MDCT), and iterative reconstruction algorithms offered stronger diagnosis and treatment planning for the benefits of patients (Hricak et al., 2011). CT scan also unlocks the door for superior treatment of some life-threatening diseases such as heart issues, strokes, cancers, and injuries due to accidents (Ginat & Gupta, 2014; Raman et al., 2013). Recently, in imaging modalities, this technology is enhancing and moving forward progressively and also landing in a successful platform due to the production of higher resolution images along with lesser scanning time required. Due to its multifaceted expansion, nowadays angiography, colonography, and urography are also covered and tested using CT scans.

CBCT is a variation of traditional CT. Unlike traditional CT scanners, CBCT data are captured with a cone-shaped x-ray beam instead of slices.

The introduction of CBCT initially happened in the field of dental implantology. But today CBCT has expanded its role in the fields of orthodontics, oral surgery, periodontics, endodontics, and pediatric dentistry and has widened its horizon in the field of forensic dentistry as well.

2.2 Benefits and uses of CT scan

CT has a profound effect on the medical field. It is particularly used for identification of substantial lesions, tumors, and metastases. It not only discloses the presence of lesions but also reveals their spatial location, size, and extent. It has the utmost role in interventional work such as minimal invasive therapy and CT-guided biopsy. Head and brain CT detect blood vessel defects and show blood clots, enlarged ventricles, and other deformities related to nerves and muscles of the eye. Thorax CT is used for detection of fibrosis, infiltration of fluids, nodular structures, and effusions (air space along with fluids). CT scan of lungs shows recurrent pulmonary infections, fibrosis, pulmonary embolism and thrombosis, calcifications, hypoplasia of lungs or artery, aortopulmonary collateral (APC) vessel malformation, pulmonary artery aneurysm, narrowing or obstructions, stenosis of the pulmonary artery or vein, poststenosis dilation, pulmonary sequestration, pulmonary neoplasm, serration, and thickening of pleura (Castañer et al., 2006). For the cardiovascular system, catheter-directed angiography and echocardiography are considered as a primary diagnostic tool; however, magnetic resonance imaging (MRI) and CT scan are also supportive imaging techniques (Tsai et al., 2008). CT quickly displays coronary heart disease (CHD)–related complications, extracardiac anomalies, and blockage; it also is helpful in surgical planning for such procedures as stent placement, vessel reimplantation, and coil embolization (Demos et al., 2004). Abdomen CT helps to diagnose kidney stones, bladder stones, internal abscesses, appendicitis, pyelonephritis, Crohn disease and ulcerative colitis of bowel, pancreatitis, liver cirrhosis, liver, kidney and spleen injuries, lymphoma, abdominal aortic aneurysms, and cancers of the kidneys, liver, pancreas, bladder, or ovaries. It helps the doctor to plan and to administer radiation therapy and helps to evaluate the response of chemotherapy on the particular tissue.

Usually doctors ask the patient to take contrast dye before doing the CT scan, and this dye can be taken orally or through intravenous route. The main objective of the administration of dye is that it makes the blood vessels and internal organs more prominent or clear on the scan.

2.3 Issues regarding CT scan

Movements of body parts are an unavoidable issue during CT scan, which leads to the production of blurred images known as artifacts. The main cause of such kind of blurred imaging is involuntary movements that include cardiac, respiratory, and gastrointestinal (Keall et al., 2006; Moorrees & Bezak, 2012). Among these three types, the respiratory organ has the largest movement. This matter of concern is solved by the introduction of various latest and efficient ways, which eradicate artifacts partially or completely from CT scan and produce a significant effect on its image. These improvements have a better impact on radiotherapy as it eliminates undesirable organ movements, leading to refined results being obtained (Karatas & Toy, 2014; Moorrees & Bezak, 2012).

Another issue related to CT scan is that it increases the risk of cancer due to radiation exposure. Ionizing radiation doses are injected in the patient during the CT scan. Useful organs of the body might develop leukemia and cancers due to long-term use of this ionizing radiation in the patient undergoing CT scan, subsequently developing the chance of cancer when exposed to high doses of ionizing radiation. Whether long-term use of low-dose ionizing radiation for diagnostic procedures can cause cancer or not is still ambiguous (Power et al., 2016).

Reaction to the contrast dye injected into the patient before the CT scan is one of the common problems encountered during or after the CT scan. Adverse effects are usually mild; however, it may be severe in some cases. Mild reactions may include nausea, vomiting, diarrhea, stomach cramp, and constipation. A potential severe reaction includes difficulty in swallowing and breathing, itching, redness of skin, hives, hoarseness, swelling of throat and airways, fast heartbeat, confusion, agitation, and peripheral cyanosis. Another issue regarding the administration of iodinated contrast material is that it can cause further kidney damage in patients who are at risk of kidney failure or have borderline kidney function. When a larger amount of dye escapes from the vein being injected, it may cause damage to the nerve, vessels, and underlying skin. Administration of contrast material should be avoided in pregnant women and mothers who are breastfeeding their babies. Recently the American College of Radiology (ACR) observed that contrast material absorbed by the baby during breastfeeding is very low (ACR, 2021). Further investigations need to be done to obtain a confirmatory conclusion.

The last issue concerning CT scan is the cost associated with this technique, and the obtained data from the CT scan may be insufficient as compared to various imaging techniques.

2.4 Technical parameters and clinical applications

Evolution of MDCT and DECT has resulted in remarkable changes in scanning parameters and radiation exposure related to CT scan. By the use of weighting algorithms, helical reconstructions, and interpolation of adjacent helical datasets, MDCT allows acquisition of various images from the same data set. Advanced musculoskeletal imaging can be obtained by using MDCT as it creates multiplanar reconstructions, maintaining high resolution. It also eliminates the chance of artifact formation, allowing scanning of the heart during a single breathhold. The main concept of using DECT to produce an image at two different energy levels (80 kVp and 140 kVp) is to gain additional information related to composition. DECT can produce a high-resolution image without increasing radiation dose as compared to single energy conventional scan. Potential applications of DECT are discussed in this section.

2.4.1 **Head and neck**

It becomes a challenge to assess inversion of laryngeal cartilage by squamous cell carcinoma by single-energy CT because uncalcified cartilage has similar attenuation to that of the tumor, but by using DECT (100 kVp and 140 kVp) potential cartilage involvement can be assessed.

2.4.2 **Lung**

Iodine has numerous disadvantages when used as contrast media. In DECT, the doctor can assess the regional distribution of ventilation by administering xenon contrast media instead of iodine. It has a superior quality of detecting pulmonary embolism and can evaluate characteristics of a pulmonary nodule as compared to conventional CT.

2.4.3 **Kidney**

DECT can be used to distinguish between a hemorrhagic cyst and a simple renal cyst. The hemorrhagic cyst appears dark on iodine display and bright on water display while simple cyst appears dark on both displays. It also helps to differentiate between small renal mass and small simple cyst. Renal mass looks isodense on water display and appears brighter than small cyst on iodine display. One of the most rising applications of DECT in the kidney is that it displays characterization of renal stone, separating them with and without uric acid (Takahashi et al., 2010). Progression of the renal lesion can also be evaluated by using dual-energy CT.

Adrenal nodules are often seen on routine abdominal CT. Further investigations are necessary to rule out possibilities of metastasis. By using DECT, characterization of adrenal nodules can be studied without any delay of further treatment, and additional health care cost can also be avoided.

2.4.4 **Liver**

DECT allows reconstruction of the virtual unenhanced image of liver parenchyma by eliminating iodine from contrast-enhanced data set, subsequently reducing radiation exposure rate to the patient. DECT detects iron load in the liver and has high accuracy in the evaluation of iron concentration even in the presence of fat. It also differentiates between small masses and cysts. Cyst looks dark in water material density image. Metastasis lesions appear less well defined or isodense to solid hepatic parenchyma on water material.

2.4.5 **Pancreas**

Without any additional unenhanced acquisition, DECT differentiates between enhancing pancreatic parenchyma and acute hemorrhage. It also helps in patient diagnosis and improves the prognosis of treatment planning as it clearly shows the areas of perfusion and necrosis in severe acute pancreatitis (Silva et al., 2011). One study showed that virtual unenhanced images can successfully be extracted using DECT from the pancreatic phase without any significant reduction in image quality, and at the same time it also reduces radiation exposure of 26.7% ± 9.7% to the patient (Mileto et al., 2011).

2.4.6 Vascular system

CT angiography can be used for diagnosis of blood vessels. Renal arteries, abdominal aortic aneurysms, circle of Willis, and carotid vessels can be rapidly imaged with minimal intervention using CT. A multiphasic CT is usually used to evaluate abdominal aortic aneurysm after endovascular repair. DECT with single-phase examination can detect aortic aneurysms with a potential decrease in radiation exposure (Stolzmann et al., 2008).

2.4.7 Heart

DECT has a role in determining stress of heart (stress test), viability imaging, and detection of cardiac iron load. According to a study done by Ruzsics et al. (2009), it was confirmed that combination of coronary CT and dual-energy cardiac perfusion is used to diagnose myocardial ischemia and coronary artery stenosis. Zhang et al. (2010) conducted a study on the canine model and confirmed DECT can detect acute myocardial infarction with 92% sensitivity and 80% specificity. DECT also helps in identifying an area of chronic myocardial infarction showing sensitivity and specificity of 77% and 97%, respectively (Bauer et al., 2010). In addition to the abovementioned advantages, DECT permits coronary CT angiography, assesses myocardial blood supply, decreases the chance of artifact formation, and reduces radiation exposure to the patient. DECT along with some software processing has the potential to analyze blood pool defects. It can identify reversible perfusion defects showing the sensitivity of 89% and specificity of 78% (Ko et al., 2011). DECT reduces the amount of false-positive results on coronary CT angiogram (CTA) and ameliorates perfection of coronary CTA (Danad et al., 2013). It can even detect coronary artery disease (CAD) in a patient with extreme calcification with high diagnostic accuracy (Scheffel et al., 2006). DECT plays an important role in the detection of cardiac iron load, thus providing benefits to the patient who cannot do MRI due claustrophobia or other contraindications (Hazirolan et al., 2008). Various authors have also investigated that DECT helps in characterization plaque, evaluation of coronary stents, and removal of calcified plaque from coronary arteries (Barreto et al., 2008; Boll et al., 2008).

2.4.8 Bone

DECT can measure marrow adipose tissue (MAT) quantities and value of marrow corrected bone marrow densities (BMD), which subsequently display mineralized tissue in a better form (Boll et al., 2008). DECT can also assess the composition of bone marrow. A more precise tool of DECT is that it can study site-specific changes in bone and those who are at high risk of fracture (e.g., in gynecologic carcinoma after chemotherapy pelvic bone is at high risk of fracture).

2.5 Advancement in CT scans

The increased clinical applications of CT examinations in all age groups has raised concerns regarding radiation dose to patients. Lowering tube current and increasing pitch have been such strategies to decrease the radiation dose (Kalra et al., 2004a). Image quality and lower radiation exposure have been reformed with the use of automatic tube-current modulation systems (Kalra et al., 2004b). CT technology has been evolving to obtain enhanced image quality, broader applications, and decreased radiation doses. Few of the developments include extreme MDCT, iterative reconstruction algorithms, DECT, phase-contrast CT, and CBCT.

2.5.1 Extreme MDCT

A multidetecting scanner consists of different patterns of scintillators, which convert x-rays into a visible spectrum light that in turn is converted by a photodiode array into waves of electric current. It also contains a shifting array that permits the shifting between channels along with the presence of a connector that relays the electrical waves to the data-acquiring portal. Previously, 64-slice CT was adequate, but currently it is not sufficient when compared to conventional catheter angiography regarding real-time visualization and anatomic details of the concerned area. Thus to attain a 0.16-mm spatial resolution provided by a fluoroscopy tube, MDCT, including 256- and 320-row detectors, was developed (Otero et al., 2009).

The investigation of the complete spectrum of complications of common cardiothoracic surgical procedures is now being efficiently carried out using MDCT angiography along with the evaluation of other noncoronary structures (CTA) (Bhatnagar et al., 2013; Nasis et al., 2013). It has also been used to quantitatively measure the size and volume of a tumor and its response during therapy (Levin et al., 2005).

The 16 cm of z axis coverage with these detectors has enhanced the visualization of the cerebrum, thereby improving the diagnosis of various ischemic diseases (Snyder et al., 2014). It has also been found that cerebrum perfusion with MDCTA yields greater accuracy in the detection of additional ischemic and partially imaged lesions (Biesbroek et al., 2013; Dorn et al., 2011). Another application of the MDCT includes the detection and evaluation of joint disease (Goh & Lau, 2012).

2.5.2 Iterative reconstruction

Iterative reconstruction algorithm essentially consists of an experimental correction loop that is launched in the image reconstruction process. Firstly, a filtered Fourier back-projection image reconstruction is accomplished in the raw data domain to create a master reconstruction. The differences observed between the measured and calculated projections are used to derive correction in the projections, its reconstruction, and is repeated until the difference becomes minor than a predetermined point. Whenever the initial image is updated, the balance in its resolution is maintained by random image processing algorithms.

2.5.3 **Dual-energy CT**

One of the following five models can be used to execute dual-energy CT:

- Dual-spin scanners: It consists of scanners that collect two data sets at two settings: low and high energy (in kVp). To collect these data, filters and the tube-current modulation can be optimized. The main disadvantage of this method is temporal misregistration between the low and high kVp scans.
- Fast kVp switching: X-ray tubes of this scanner rapidly shift between high- and low-voltage settings along with an ultrafast detector, which is necessary to obtain multiple projections in every rotation. The main disadvantages of this fast shifting are the overlap between the two energy spectra, difficulty in selectively interposing a filter in the high-energy exposure, and a slightly increased radiation dose.
- Dual-source scanners: The scanners in this have two independent imaging chains mounted on a CT gantry, which operate in a low-energy and a high-energy mode that allows simultaneous recording of two independent data sets, at both energies. Optimization of the imaging chain can be done individually, as well as modulation of tube current if deemed necessary. As the two imaging chains work simultaneously, contamination from one scatter to the other can occur.
- Dual-layer detectors: This consists of an x-ray source that is polychromatic, hence it enables acquiring a low- and high-energy spectral band from one exposure thus allowing synchronous projection. It provides immunity from cross scatter, projection-level dual processing, and modulation of tube current for dose optimization. An advantage of this model is that dual-energy processing can be accomplished retrospectively.
- Photon-counting detectors: These detectors separate incoming photons into two different compartments according to the photon energy. Around two to eight compartments may be used to allow decomposition of every projection into multiple spectral bands. These detectors demonstrate remarkable noise and dose reduction, improved contrast, and multiple-energy imaging.

Regardless of the method for acquiring different energy bands, the postprocessing steps for the abovementioned five models are quite similar. With every increasing x-ray photon energy, the total attenuation reduces, and this reduction is characteristic of the material composition of each voxel. Some of the drawbacks of DECT include a slight increase in radiation dose, small field of view (FOV with dual-source CT), and noisy low-energy images (80 kVp).

2.5.4 Phase-contrast CT

In a phase-contrast CT, instead of x-ray attenuation, the phase shift of x-rays passes through the matter to generate contrast in the tissue. It is found to yield appreciable contrast in biologic soft tissue than absorption-based x-ray imaging. The information gathered can be obtained through various methods such as analyzer based, crystal interferometer based, or propagation based. Majority of the methods rely on monochromatic, parallel x-ray beams, and therefore x-ray sources such as synchrotron radiation or low-power microfocus x-ray tubes pose a problem in a clinical scenario (Hetterich et al., 2014).

2.5.5 Cone beam CT

An x-ray beam shaped in the form of a cone, centered on a two-dimensional (2D) detector, rotating around the object, produces several, sequential, and planar images that are reconstructed into a volumetric data file. Scanning of the intended area is performed in one rotation thereby notably decreasing the radiation exposure.

The CBCT scanners use a 2D digital array, which provides area detector rather than linear detector, and the voxel resolutions are equal in all three dimensions. The operating range of a CBCT x-ray is between 1 and 15 mA at 90 to 120 kVp, whereas that of a CT is 120 to 150 mA at 220 kV. The scanned images are relayed to the system to reconstruct it into an anatomic volume for viewing in axial, sagittal, and coronal planes using a modified Feldkamp algorithm.

2.6 Application of CBCT in dentistry
2.6.1 CBCT in dental implantology

CBCT has transformed the way dental treatment is performed in dental establishments. The overall benefit of the implant therapy with less surgical and postoperative complications is enhanced only if the implantologist has three-dimensional (3D) information about bone volume and topography before placement of the implant (Tyndall & Rathore, 2008). The imaging technique guides through the accurate assessment of bone volume, bone density, and proximity to the anatomic structures in the implant site. CBCT imparts the benefit of showing the different types of alveolar ridge pattern such as an irregular ridge, narrow crestal ridge, or knife shape ridge (Jaju & Jaju, 2013). Bone density obtained in CT units is expressed in terms of Hounsfield units [HU]. CBCT determines the type and site of the planned implant, its position within the bone, its association to the planned restoration and adjacent teeth/implants, and proximity to the vital structure that can be achieved with the assimilation of the CBCT scans with computer-aided design/computer-aided manufacturing technology (e.g., CEREC; Sirona Dental GmbH, Salzburg, Austria). Computer-generated surgical guides can be prepared from the virtual treatment plan, which is used to place the planned implants in the patient's mouth in the exact position determined, thus allowing more predictable implant placement (Orentlicher & Abboud, 2011).

The use of computer-guided implant surgical guides are of the following types:

1. Tooth supported
2. Mucosa supported
3. Bone supported

Tooth-supported guides are preferred in partially edentulous cases. Mucosa-supported guides are used in fully edentulous cases. Bone-supported guides are used in partially/fully edentulous cases in which remarkable ridge atrophy is a present and good adaptation of a mucosa-supported guide is questionable.

Currently, only SimPlant (Materialise Dental, Leuven, Belgium) produces bone-supporting surgical guides (Hara, 2014).

American Academy of Oral and Maxillofacial Radiology (AAOMR) guidelines for the role of CBCT in dental implantology.

Recommendation 1: The imaging modality of choice in the initial evaluation of a dental implant patient is usually panoramic radiography.

Recommendation 2: Intraoral radiography should be used to supplement the preliminary information from panoramic radiography.

Recommendation 3: CBCT shouldn't be used as an initial diagnostic tool for imaging examination.

Recommendation 4: The radiographic examination of an implant site should be cross-sectional imaging orthogonal to the site of interest.

Recommendation 5: CBCT should be used as the imaging modality of choice for preoperative imaging of potential implant sites.

Recommendation 6: Clinical conditions indicating the need for augmentation procedures or site development before the placement of dental implants are (1) sinus augmentation, (2) block or particulate bone grafting, (3) ramus or symphysis grafting, (4) assessment of impacted teeth in the field of interest, and (5) evaluation of prior traumatic injury require CBCT.

Recommendation 7: Surgical procedures such as bone reconstruction and augmentation procedures (e.g., ridge preservation or bone grafting) have been performed to treat bone volume deficiencies before implant placement require CBCT imaging.

Recommendation 8: In the absence of clinical signs or symptoms, intraoral periapical radiography is used for the postoperative assessment of implants. Panoramic radiographs may be indicated for more extensive implant therapy cases.

Recommendation 9: It is advisable to use cross-sectional imaging [CBCT] postoperatively immediately only if the patient presents with impaired mobility and altered sensation especially if the fixture is in the posterior mandible.

Recommendation 10: For the periodic review of clinically asymptomatic implants it is not advisable to use CBCT.

Recommendation 11: Usually CBCT is advised when implant retrieval is to be done.

2.6.2 CBCT in oral and maxillofacial surgery

An amalgamation of high-quality bone definition, low radiation dose, and compact design requiring minimum space has made CBCT desirable for the evaluation and examination of pathologies in the head and the neck as well as in extracranial, paranasal, and temporal regions (Adibi et al., 2012).

Third molar evaluation

There is a significant improvement in the poor visualization of the mandibular canal seen in the panoramic images (Kamrun et al., 2013).

CBCT is more reliable than panoramic radiograph in determining the number of roots (Suomalainen et al., 2010).

The CBCT software has a nerve tracing application that shows the color coding and identification of the mandibular nerve.

The position of the impacted canine varies greatly, whereas CBCT with its 3D orientation can specifically locate its position, thus assisting the oral surgeon in the proper treatment plan (Marques et al., 2010).

Bony pathology assessments

CBCT is preferred as the imaging modality when there is a need to diagnose cyst, tumor, and infections in the alveolar process and jawbone. Rare calcifying lesions such as a calcifying cystic odontogenic tumor can be evaluated properly in CBCT images for their particular variations. CBCT plays a very important role in the evaluation of intraosseous lesions that are near vital organs and vasculature in the head and neck region. The combination of dynamic contrast-enhanced MRI and CBCT can be used in defining tumor boundaries and in developing appropriate surgical interventions (Hendrikx et al., 2010). CBCT collaborated with stereolithographic model construction helps in dental implant placement or the reconstruction of jaws removed due to pathology (Guttenberg, 2008).

Maxillofacial trauma

The certitude for diagnosis becomes higher for surgeons who use CBCT imaging than conventional radiography (Kaeppler et al., 2013). CBCT is used as a confirmatory imaging modality in maxillofacial trauma. Fractures such as a dentoalveolar fracture, incomplete cortical plate fracture, gunshot injury, maxillary bone fracture, zygomatic complex fracture, and a mandibular fracture require imaging techniques for knowing the exact location of the fracture.

Bone graft analysis

The volumetric analysis gives a better picture of defect morphology as in the case of cleft palate. The morphology of a traumatic defect is very difficult to understand, which is of utmost importance before placement of the implant. Defect size and shape can alter the factors that guide the treatment planning decisions.

Temporomandibular joint (TMJ) assessment

Recent study limitations in mandibular movement and function, stiffness in jaw and pain in the TMJ, and improper visualization of the articular eminence in panoramic radiograph, suggest the usage of CBCT as an adjunct to other methods of imaging (de Boer et al., 2014).

Craniofacial surgery

Dental age, arch segment positioning, and cleft size can be better assessed using CBCT (Quereshy et al., 2008).

Orthognathic surgery

Certain CBCT machines (e.g., GALILEOS [Sirona Dental GmbH Bensheim, Germany]) have an inbuilt face scanner, which helps in depicting a virtual mirror image of the patient to assist in patient education and surgical treatment planning.

A 3D model can be virtually created using CBCT and used for orthodontic and orthognathic analysis thereby assessing the surgical protocol and final prognosis (Quereshy et al., 2008).

2.6.3 **CBCT in endodontics**

CBCT exhibited the periodontal ligament space more accurately, thus serving in all stages in endodontics (Jervøe-Storm et al., 2010).

The 3D Accuitomo (J Morita USA, Inc.), the first of the small FOV systems, showed a resolution of 0.125 mm (Jaju, 2013).

Assessment of root canal morphology

The prevalence of a second mesiobuccal canal [MB2] in the maxillary first molars has been varying from 69% to 93%, and conventional radiographs have reported only 55% of this variation in configuration (Ramamurthy et al., 2006).

Ramamurthy et al. (2006) and Matherne et al. (2008) have highlighted the limitations of 2D imaging in the detection of the MB2 canal.

CBCT has shown the high prevalence of the distolingual canal in Taiwanese individuals, it highlights the defects in the root canal system of mandibular premolars, and it helps in the determination of root curvature (Kleghorn et al., 2008; Tu et al., 2009).

Some authors have also reported the anatomic variations in mandibular incisor detected in an Indian population by CBCT (Jaju, 2013; Jaju et al., 2013).

Dental periapical pathoses

Lesions extending only to the cancellous bone with little or no cortical plate involvement becomes tricky to diagnose with the intraoral film.

Apical periodontitis can be detected at a higher rate using CBCT rather than periapical and panoramic radiograph (Estrela et al., 2008).

Estrela et al. (2008) proposed a periapical index based on CBCT for the identification of apical periodontitis (AP). The CBCT periapical index is a six-point (0–5) scoring system calculated by assessing the largest lesional measurement in either the buccopalatal, mesiodistal, or diagonal dimension, and it also considers expansion and destruction of cortical bone.

Low et al. (2008) concluded that CBCT worked better in the detection of periapical lesions compared with conventional imaging. CBCT technology and the assignment of grey values helps to differentiate between cysts and granulomas (Jaju, 2013).

Root resorption

CBCT has also been applied in the assessment of postorthodontic root resorption, especially the roots of maxillary lateral incisors impacted by maxillary canine (Jaju et al., 2013). The confirmation of internal root resorption and its difference from external root resorption has been well established by CBCT (Cohenca et al., 2007)

The treatment complexity can be well assessed by CBCT based on the extent of the resorptive lesion, and this helps the clinician in proper prognosis thereby making both treatment and treatment outcomes more predictable (Jaju et al., 2013).

Postoperative assessment

CBCT is used in the initial and subsequent monitoring of periapical lesions for proper healing. Proximity to the mandibular canal, mental foramen, and maxillary sinus during periapical surgery can be precisely assessed on CBCT images.

Recommendations of the American Association of Endodontists (AAE) and American Association of Oral and Maxillofacial Radiology (AAOMR)

The following are recommendations from the AAE and AAOMR (2010)

1. Identification of accessory canals in teeth with different complex morphology according to conventional imaging.
2. Identification of root canal defects and assessment of root curvature.
3. Diagnosis of periapical pathosis in patients with inconsistent or contrasting clinical signs and symptoms such as an untreated or a previously treated tooth with less localized symptoms showing no features of pathosis as already identified by conventional imaging techniques and in cases of anatomic superimposition of roots or other areas of maxillofacial surface requiring operative procedures.
4. Diagnosis of nonendodontic lesion, its extent, and its effect on surrounding structures.
5. Intra- or postoperative assessment of procedural error and complications such as overextrusion of root canal obturation material, fractured endodontic instruments, calcified canal identification, and locating perforations.
6. Diagnosis and management of dentoalveolar trauma such as dentoalveolar fractures, root fractures, and luxation.
7. Localization and differentiation of external and internal root resorption or invasive cervical resorption from other conditions, its definitive treatment plan, and prognosis.
8. Presurgical treatment plan that helps to locate the root apex/apices and evaluation of its proximity toward close or adjacent anatomic landmarks.
9. Dental implant case planning, which is based on the evaluation of the edentulous ridge 3D imaging, becomes the most important approach.

Operative dentistry

CBCT helps in the detection of proximal caries but not occlusal caries. Its effect was different in deep enamel, superficial dentin, and deep dentin.

2.6.4 **Applications in orthodontics**

Launch of new software in orthodontic assessment, such as Dolphin (Dolphin Imaging and Management Solutions, Chatsworth, CA, USA) and In vivo Dental (Anatomage, San Jose, CA, USA) has enabled dentists to use CBCT images for cephalometric analysis, facial growth, age, airway function, and disturbances in tooth eruption (Bjerklin & Ericson, 2006).

A study done by Moreira et al. (2009) has substantiated the accuracy of linear measurements from CBCT cephalograms, which proved to be exact than the lateral cephalogram by comparing skull measurements.

The proximity of impacted teeth to vital structures interfering with orthodontic movement can be assessed by CBCT (Erickson et al., 2003).

CBCT is a reliable tool for safe insertion of mini-implants required as temporary anchors, thus preventing procedural and irreparable damage to the roots (Alamri et al., 2012).

CBCT is used to evaluate bone density before, during, and after treatment (Alamri et al., 2012).

CBCT can be used in different aspects of orthodontic treatment. According to Kapila and Nervina (2015)**, CBCT should be preferred because:**

- It helps in the location of the impacted canine and supernumerary teeth.
- It helps to gauge the enormity of the defect in patients with Cleft Lip and Palate (CLP).
- It modifies the differential diagnosis of patients with malocclusions such as craniofacial anomalies and syndromes.
- It helps to identify the discrepancy whether it is unilateral or bilateral for facial asymmetry mainly for patients with orthognathic surgery.
- It recognizes the etiology of the malocclusion such as TMJ disorders.
- It plays an important role in the outcome of treatment such as rapid maxillary expansion and root angulations.
- It assesses the quality and quantity of bone and the anatomic structures when orthodontic device placements such as miniscrews are needed.
- It evaluates the condition of the alveolar boundary.
- It is also used for obstructive sleep apnea and helps to visualize the 3D array morphology.

2.6.5 **Pediatric dentistry**

Basic principles of radiation protection

1. Justification principle. Radiographs should be used only to obtain necessary information, which by any means is not obtainable. If the patient finds it difficult to cope with the procedure, then radiographs are not advisable.

2. Limitation principle. It states that the radiation dose should be maintained as low as reasonably achievable (ALARA) as supported by the American Dental Association.
3. Optimization principle. It states that any practitioner should always try to achieve the best possible diagnostic image.

Indications
Caries
CBCT images help in better detection of proximal carious lesions, but it is unable to detect carious lesions beneath metal restored crowns and tooth with radiopaque restorations.

Diagnosis of supernumerary teeth
CBCT evaluation of impacted supernumerary teeth is recommended to reduce the risk of damage to the surrounding anatomic structures as they are in close association with the cortical bone.

Endodontic applications
Promising diagnostic tool for complex endodontic cases. It shows the extent of periapical pathologies, perforations, obturations, root fractures, and location of fractured root canal instruments in root canals. CBCT gives an enhanced view of calcified canals and missed canals and helps to measure root length and angle of curvature.

Dental trauma
Root fractures of nonendodontically treated teeth are better visualized in CBCT.

TMJ disorders
These are more advanced in the assessment of osseous TMJ abnormalities.

Orthodontic treatment
Valuable diagnostic information can be achieved in pediatric patients undergoing orthodontic treatment through CBCT. CBCT helps in the assessment of ankylosed and submerged primary tooth, evaluation of impacted canine and premolar, evaluation of buccal and lingual cortical plates, and assessment of proposed sites of temporary anchorage devices. It acts as an adjunct to orthognathic surgeries.

Forensic odontology
Currently, CBCT is also used in forensic odontology for age estimation, forensic facial reconstruction, and analysis of bitemarks, sex determination, and frontal sinus pattern.

2.6.6 Periodontics
Different intrabony defects such as dehiscence, fenestration defects, and periodontal cysts can be measured precisely with the help of CBCT. CBCT is also used to locate the furcation involvement of periodontal defects, which can help the clinician to assess the postsurgical results of regenerative periodontal therapy (Eshraghi et al., 2019).

Authors have evaluated that CBCT has higher accuracy in determining the periodontal ligament space than conventional radiographs (Jervøe-Storm et al., 2010b).

Assessment of bone grafts and periodontal regenerative therapy outcomes is better and more accurate with CBCT imaging (Eshraghi et al., 2012; Grimard et al., 2009).

2.6.7 Forensic dentistry

Age estimation can be assessed by the pulp-dentin complex, which shows morphologic changes not always possible by extraction and sectioning the tooth. Hence CBCT provides a noninvasive alternative (Yang et al., 2021).

Skeletal age assessment can be done by visualization of cervical vertebral morphology, and CBCT volumetric data sets help in proper segmentation of individual vertebrae, which provides a 3D method to the biologic aging of orthodontic patients with the help of images of the cervical spine. It also aids in studying disease processes such as spinal fractures consequent to osteoporosis (Shi et al., 2007).

CBCT is used in forensic investigations for postmortem imaging and 3D imaging in case of gunshot injury as it detects structural hard tissue damage by a high-density metal projectile and determines the exact location of the projectile in the body (von See et al., 2009).

2.7 Conclusion

With the advent of CT imaging, the identification of acquired and congenital malformation has ameliorated the diagnosis. The high-resolution images obtained through micro-CT and CBCT not only refine treatment plan but can also be used by the researchers to document the effects of surgical intervention and its outcomes as well as to formulate and evaluate different hypotheses. It has revolutionized radiology, but because of its widespread use in all age groups it still carries the drawback of high radiation exposure. Hence development in hardware and software systems of CT will help overcome this issue thereby increasing its use in clinical practice.

References

Adibi, S., Zhang, W., Servos, T., O'Neill, P., 2012. Cone beam computed tomography for general dentists. Open Access Scientific Reports, 1.

Alamri, H.M., Sadrameli, M., Alshalhoob, M.A., Sadrameli, M., Alshehri, M.A., 2012. Applications of CBCT in dental practice: A review of the literature. General Dentistry 60, 390–400.

American Association of Endodontists and American Academy of Oral and Maxillofacial Radiology, 2010. Joint position statement of the American Association of Endodontists and the American Academy of Oral and Maxillofacial Radiology: Use of cone-beam computed tomography in endodontics. www.aaomr.org/assets/Journal_Publishers/Position_Papers/aaomr-aac_position_paper_cb_pdf.

American College of Radiology, 2021. ACR manual on contrast media. ACR Committee on Drugs and Contrast Media. https://www.acr.org/Clinical-Resources/Contrast-Manual.

Barreto, M., Schoenhagen, P., Nair, A., Amatangelo, S., Milite, M., Obuchowski, N.A., Lieber, M.L., Halliburton, S.S., 2008. Potential of dual-energy computed tomography to characterize atherosclerotic plaque: Ex vivo assessment of human coronary arteries in comparison to histology. Journal of Cardiovascular Computed Tomography 2, 234–242. https://doi.org/10.1016/j.jcct.2008.05.146.

Bauer, R.W., Kerl, J.M., Fischer, N., Burkhard, T., Larson, M.C., Ackermann, H., Vogl, T.J., 2010. Dual-energy CT for the assessment of chronic myocardial infarction in patients with chronic coronary artery disease: Comparison with 3-T MRI. American Journal of Roentgenology 195, 639–646. https://doi.org/10.2214/AJR.09.3849.

Bhatnagar, G., Vardhanabhuti, V., Nensey, R.R., Sidhu, H.S., Morgan-Hughes, G., Roobottom, C.A., 2013. The role of multidetector computed tomography coronary angiography in imaging complications post-cardiac surgery. Clinical Radiology 68, e254–e265. https://doi.org/10.1016/j.crad.2012.11.015.

Biesbroek, J.M., Niesten, J.M., Dankbaar, J.W., Biessels, G.J., Velthuis, B.K., Reitsma, J.B., Schaaf, I.C., 2013. Diagnostic accuracy of CT perfusion imaging for detecting acute ischemic stroke: A systematic review and meta-analysis. CED 35, 493–501. https://doi.org/10.1159/000350200.

Bjerklin, K., Ericson, S., 2006. How a computerized tomography examination changed the treatment plans of 80 children with retained and ectopically positioned maxillary canines. The Angle Orthodontist 76, 43–51.

Boll, D.T., Merkle, E.M., Paulson, E.K., Mirza, R.A., Fleiter, T.R., 2008. Calcified vascular plaque specimens: Assessment with cardiac dual-energy multidetector CT in anthropomorphically moving heart phantom. Radiology 249, 119–126. https://doi.org/10.1148/radiol.2483071576.

Castañer, E., Gallardo, X., Rimola, J., Pallardó, Y., Mata, J.M., Perendreu, J., Martin, C., Gil, D., 2006. Congenital and acquired pulmonary artery anomalies in the adult: Radiologic overview. Radiographics 26, 349–371. https://doi.org/10.1148/rg.262055092.

Cleghorn, B.M., Christie, W.H., Dong, C.C.S., 2008. Anomalous mandibular premolars: A mandibular first premolar with three roots and a mandibular second premolar with a C-shaped canal system. International Endodontic Journal 41, 1005–1014.

Cohenca, N., Simon, J.H., Mathur, A., Malfaz, J.M., 2007. Clinical indications for digital imaging in dento-alveolar trauma. Part 2: Root resorption. Dental Traumatology 23, 105–113.

Danad, I., Raijmakers, P.G., Knaapen, P., 2013. Diagnosing coronary artery disease with hybrid PET/CT: It takes two to tango. Journal of Nuclear Cardiology 20, 874–890. https://doi.org/10.1007/s12350-013-9753-8.

de Boer, E.W., Dijkstra, P.U., Stegenga, B., de Bont, L.G., Spijkervet, F.K., 2014. Value of cone-beam computed tomography in the process of diagnosis and management of disorders of the temporomandibular joint. British Journal of Oral and Maxillofacial Surgery 52, 241–246.

Demos, T.C., Posniak, H.V., Pierce, K.L., Olson, M.C., Muscato, M., 2004. Venous anomalies of the thorax. American Journal of Roentgenology 182, 1139–1150. https://doi.org/10.2214/ajr.182.5.1821139.

Dorn, F., Muenzel, D., Meier, R., Poppert, H., Rummeny, E.J., Huber, A., 2011. Brain perfusion CT for acute stroke using a 256-slice CT: Improvement of diagnostic information by large volume coverage. European Radiology 21, 1803–1810. https://doi.org/10.1007/s00330-011-2128-0.

Erickson, M., Caruso, J.M., Leggitt, L., 2003. Newtom QR-DVT 9000 imaging used to confirm a clinical diagnosis of iatrogenic mandibular nerve paresthesia. Journal of the California Dental Association 31, 843–845.

Eshraghi, T., McAllister, N., McAllister, B., 2012. Clinical applications of digital 2-D and 3-D radiography for the periodontist. Journal of Evidence Based Dental Practice 12, 36–45. 23040338. https://doi.org/10.1016/S1532-3382(12)70010-6.

Eshraghi, V.T., Malloy, K.A., Tahmasbi, M., 2019. Role of cone-beam computed tomography in the management of periodontal disease. Dentistry Journal (Basel), 7.

Estrela, C., Bueno, M.R., Azevedo, B.C., Azevedo, P., et al., 2008. A new periapical index based on cone beam computed tomography. Journal of Endodontics 34, 1325–1331.

Ginat, D.T., Gupta, R., 2014. Advances in computed tomography imaging technology. Annual Review of Biomedical Engineering 16, 431–453. https://doi.org/10.1146/annurev-bioeng-121813-113601.

Goh, Y.P., Lau, K.K., 2012. Using the 320-multidetector computed tomography scanner for four-dimensional functional assessment of the elbow joint. American Journal of Orthopedics (Belle Mead NJ) 41, e20–e24.

Grimard, B.A., Hoidal, M.J., Mills, M.P., Mellonig, J.T., Nummikoski, P.V., Mealey, B.L., 2009. Comparison of clinical, periapical radiograph, and cone-beam volume tomography measurement techniques for assessing bone level changes following regenerative periodontal therapy. Journal of Periodontology 80, 48–55.

Guttenberg, S.A., 2008. Oral and maxillofacial pathology in three dimensions. Dental Clinics of North America 52, 843–873.

Hara, S., 2014. Computer-aided design provisionalization and implant insertion combined with optical scanning of plaster casts and computed tomography data. Annals of Maxillofacial Surgery 4 (1), 64–69.

Hazirolan, T., Akpinar, B., Unal, S., Gümrük, F., Haliloglu, M., Alibek, S., 2008. Value of Dual Energy Computed Tomography for detection of myocardial iron deposition in Thalassaemia patients: Initial experience. European Journal of Radiology 1872–7727, 68 (3), 442–445. doi:10.1016/j.ejrad.2008.07.014. PMID: 18768275.

Hendrikx, A.W., Maal, T., Dieleman, F., Van Cann, E.M., Merkx, M.A., 2010. Cone-beam CT in the assessment of mandibular invasion by oral squamous cell carcinoma: Results of the preliminary study. International Journal of Oral and Maxillofacial Surgery 39, 436–439.

Hetterich, H., Willner, M., Fill, S., Herzen, J., Bamberg, F., Hipp, A., Schüller, U., Adam-Neumair, S., Wirth, S., Reiser, M., Pfeiffer, F., Saam, T., 2014. Phase-contrast CT: Qualitative and quantitative evaluation of atherosclerotic carotid artery plaque. Radiology 271, 870–878. https://doi.org/10.1148/radiol.14131554.

Hricak, H., Adelstein, S.J., Hall, E.J., Howell, R.W., Mccollough, C.H., Mettler, F.A., Pearce, M.S., Thrall, J.H., Wagner, L.K., 2011Managing Radiation Use in Medical Imaging: A Multifaceted Challenge258, 889–905.

Jaju, P. P., & Jaju, S. P. (2013). Suvarna PV. In P. Dedhia (Ed.), *Dental CT: Third eye in dental implants*. Jaypee Brothers Medical Publishers Ltd.

Jaju, S.P., 2013. Cone beam CT in endodontics: A paradigm shift in clinical practice. Smile Dental Journal 8, 22–28.

Jaju, S.P., Jaju, P.P., Garcha, V., 2013. Root canal assessment of mandibular incisors in an Indian population using cone beam CT. Endodontics Practice 7, 105–11122.

Jervøe-Storm, P.M., Hagner, M., Neugebauer, J., 2010. Comparison of cone-beam computerized tomography and intraoral radiographs for determination of the periodontal ligament in a variable phantom. Oral Surgery, Oral Medicine, Oral Pathology, Oral Radiology, and Endodontology 109, 95–101.

Kaeppler, G., Cornelius, C.P., Ehrenfeld, M., Mast, G., 2013. Diagnostic efficacy of cone-beam computed tomography for mandibular fractures. Oral Surgery, Oral Medicine, Oral Pathology, Oral Radiology 116, 98–14.

Kalra, M.K., Maher, M.M., Toth, T.L., Hamberg, L.M., Blake, M.A., Shepard, J.-A., Saini, S., 2004a. Strategies for CT radiation dose optimization. Radiology 230, 619–628. https://doi.org/10.1148/radiol.2303021726.

Kalra, M.K., Maher, M.M., Toth, T.L., Schmidt, B., Westerman, B.L., Morgan, H.T., Saini, S., 2004b. Techniques and applications of automatic tube current modulation for CT. Radiology 233, 649–657. https://doi.org/10.1148/radiol.2333031150.

Kamrun, N., Tetsumura, A., Nomura, Y., 2013. Visualization of the superior and inferior borders of the mandibular canal: a comparative study using digital panoramic radiographs and cross-sectional computed tomography images. Oral Surgery, Oral Medicine, Oral Pathology, Oral Radiology 115, 550–557.

Kapila, S.D., Nervina, J.M., 2015. CBCT in orthodontics: Assessment of treatment outcomes and indications for its use. Dentomaxillofacial Radiology 44, 20140282.

Karatas, O.H., Toy, E., 2014. Three-dimensional imaging techniques: A literature review. European Journal of Dentistry 8, 132–140. https://doi.org/10.4103/1305-7456.126269.

Keall, P.J., Mageras, G.S., Balter, J.M., Emery, R.S., Forster, K.M., Jiang, S.B., Kapatoes, J.M., Low, D.A., Murphy, M.J., Murray, B.R., Ramsey, C.R., Van Herk, M.B., Vedam, S.S., Wong, J.W., Yorke, E., 2006. The management of respiratory motion in radiation oncology report of AAPM Task Group 76. Medical Physics 33, 3874–3900. https://doi.org/10.1118/1.2349696.

Ko, S.M., Choi, J.W., Song, M.G., Shin, J.K., Chee, H.K., Chung, H.W., Kim, D.H., 2011. Myocardial perfusion imaging using adenosine-induced stress dual-energy computed tomography of the heart: Comparison with cardiac magnetic resonance imaging and conventional coronary angiography. European Radiology 21, 26–35. https://doi.org/10.1007/s00330-010-1897-1.

Levin, D.C., Rao, V.M., Parker, L., Frangos, A.J., Sunshine, J.H., 2005. Recent trends in utilization of cardiovascular imaging: How important are they for radiology? Journal of the American College of Radiology 2, 736–739. https://doi.org/10.1016/j.jacr.2005.01.015.

Low, K.M., Dula, K., Bürgin, W., Arx, T., 2008. Comparison of periapical radiography and limited cone-beam tomography in posterior maxillary teeth referred for apical surgery. Journal of Endodontics 34, 557–562.

Marques, Y.M., Botelho, T.D., Xavier, F.C., Rangel, A.L., Rege, I.C., 2010. Importance of cone beam computed tomography for diagnosis of calcifying cystic odontogenic tumour associated to odontoma. Report of a case. Medicina Oral, Patologia Oral, Cirugia Bucal, 15.

Matherne, R.P., Angelopoulos, C., Kulild, J.C., Tira, D., 2008. Use of cone-beam computed tomography to identify root canal systems in vitro. Journal of Endodontics 34, 87–89.

Mileto, A., Mazziotti, S., Gaeta, M., Bottari, A., Zimbaro, F., Giardina, C., Ascenti, G., 2011. Pancreatic dual-source dual-energy CT: Is it time to discard unenhanced imaging? Clinical Radiology 67, 334–339. https://doi.org/10.1016/j.crad.2011.09.004.

Moorrees, J., Bezak, E., 2012. Four dimensional radiotherapy: A review of current technologies and modalities. Australasian Physical & Engineering Sciences 35. Medicine. Supported by the Australasian College of Physical Scientists in Medicine and the Australasian Association of Physical Sciences in Medicine. https://doi.org/10.1007/s13246-012-0178-5.

Moreira, C.R., Sales, M.A., Lopes, P.M., Cavalcanti, M.G., 2009. Assessment of linear and angular measurements on three-dimensional cone-beam computed tomographic images.

Oral Surgery, Oral Medicine, Oral Pathology, Oral Radiology, and Endodontics 108, 430–436.

Nasis, A., Mottram, P.M., Cameron, J.D., Seneviratne, S.K., 2013. Current and evolving clinical applications of multidetector cardiac CT in assessment of structural heart disease. Radiology 267, 11–25. https://doi.org/10.1148/radiol.13111196.

Orentlicher, G., Abboud, M., 2011. Guided surgery for implant therapy. Dental Clinics of North America 55, 715–744.

Otero, H.J., Steigner, M.L., Rybicki, F.J., 2009. The "post-64" era of coronary CT angiography: Understanding new technology from physical principles. Radiology Clinics of North America 47, 79–90. https://doi.org/10.1016/j.rcl.2008.11.001.

Power, S.P., Moloney, F., Twomey, M., James, K., O'Connor, O.J., Maher, M.M., 2016. Computed tomography and patient risk: Facts, perceptions and uncertainties. World Journal of Radiology 1949–8470, 8 (12), 902–915. doi:10.4329/wjr.v8.i12.902. PMID: 28070242.

Quereshy, F.A., Savell, T.A., Palomo, J.M., 2008. Applications of cone beam computed tomography in the practice of oral and maxillofacial surgery. Journal of Oral and Maxillofacial Surgery 66, 791–796.

Ramamurthy, R., Scheetz, J.P., Clark, S.J., Farman, A.G., 2006. Effects of imaging system and exposure on accurate detection of the second mesio-buccal canal in maxillary molar teeth. Oral Surgery, Oral Medicine, Oral Pathology, Oral Radiology, and Endodontics 102, 796–802.

Raman, S.P., Mahesh, M., Blasko, R.V., Fishman, E.K., 2013. CT scan parameters and radiation dose: Practical advice for radiologists. Journal of the American College of Radiology 10, 840–846. https://doi.org/10.1016/j.jacr.2013.05.032.

Ruzsics, B., Schwarz, F., Schoepf, U.J., Lee, Y.S., Bastarrika, G., Chiaramida, S.A., Costello, P., Zwerner, P.L., 2009. Comparison of dual-energy computed tomography of the heart with single photon emission computed tomography for assessment of coronary artery stenosis and of the myocardial blood supply. American Journal of Cardiology 104, 318–326. https://doi.org/10.1016/j.amjcard.2009.03.051.

Scheffel, H., Alkadhi, H., Plass, A., Vachenauer, R., Desbiolles, L., Gaemperli, O., Schepis, T., Frauenfelder, T., Schertler, T., Husmann, L., Grunenfelder, J., Genoni, M., Kaufmann, P.A., Marincek, B., Leschka, S., 2006. Accuracy of dual-source CT coronary angiography: First experience in a high pre-test probability population without heart rate control. European Radiology 16, 2739–2747. https://doi.org/10.1007/s00330-006-0474-0.

Shi, H., Scarfe, W.C., Farman, A.G., 2007. Three-dimensional reconstruction of individual cervical vertebrae from cone-beam computed-tomography images. American Journal of Orthodontics and Dentofacial Orthopedics, 131.

Silva, A.C., Morse, B.G., Hara, A.K., Paden, R.G., Hongo, N., Pavlicek, W., 2011. Dual-energy (spectral) CT: Applications in abdominal imaging. Radiographics 31, 1031–1046. discussion 1047–1050. https://doi.org/10.1148/rg.314105159.

Snyder, K.V., Mokin, M., Bates, V.E., 2014. Neurologic applications of whole-brain volumetric multidetector computed tomography. Neurology Clinics 32, 237–251. https://doi.org/10.1016/j.ncl.2013.08.001.

Stolzmann, P., Frauenfelder, T., Pfammatter, T., Peter, N., Scheffel, H., Lachat, M., Schmidt, B., Marincek, B., Alkadhi, H., Schertler, T., 2008. Endoleaks after endovascular abdominal aortic aneurysm repair: Detection with dual-energy dual-source CT. Radiology 249, 682–691. https://doi.org/10.1148/radiol.2483080193.

Suomalainen, A., Ventä, I., Mattila, M., Turtola, L., Vehmas, T., Peltola, J.S., 2010. Reliability of CBCT and other radiographic methods in preoperative evaluation of lower third molars. Oral Surgery, Oral Medicine, Oral Pathology, Oral Radiology, and Endodontics 109, 276–284.

Takahashi, N., Vrtiska, T.J., Kawashima, A., Hartman, R.P., Primak, A.N., Fletcher, J.G., McCollough, C.H., 2010. Detectability of urinary stones on virtual nonenhanced images generated at pyelographic-phase dual-energy CT. Radiology 256, 184–190. https://doi.org/10.1148/radiol.10091411.

Tsai, I-C., Chen, M.-C., Jan, S.-L., Wang, C.-C., Fu, Y.-C., Lin, P.-C., Lee, T., 2008. Neonatal cardiac multidetector row CT: Why and how we do it. Pediatric Radiology 0301–0449, 38 (4), 438–451. doi:10.1007/s00247-008-0761-9. PMID: 18259739.

Tu, M.G., Huang, H.-L., Hsue, S.-S., Hsu, J.-T., Chen, S.-Y., Jou, M.-J., Tsai, C.-C., 2009. Detection of permanent three-rooted mandibular first molars by cone-beam computed tomography imaging in Taiwanese individuals. Journal of Endodontics 35, 503–507.

Tyndall, D.A., Rathore, S., 2008. Cone-beam CT diagnostic applications: Caries, periodontal bone assessment, and endodontic applications. Dental Clinics of North America 52, 825–841.

von See, C., Bormann, K.-H., Schumann, P., Goetz, F., Gellrich, N.-C., Rücker, M.S., 2009. Forensic imaging of projectiles using cone-beam computed tomography. Forensic Science International 190, 38–41.

Yang, F., Jacobs, R., Willems, G., 2021. Dental age estimation through volume matching of teeth imaged by cone-beam CT. Forensic Science International 159 (1), S78–S83.

Zhang, L.-J., Peng, J., Wu, S.-Y., Yeh, B.M., Zhou, C.-S., Lu, G.-M., 2010. Dual source dual-energy computed tomography of acute myocardial infarction: Correlation with histopathologic findings in a canine model. Investigative Radiology 45, 290–297. https://doi.org/10.1097/RLI.0b013e3181dfda60.

Ultrasonography Technology and applications in clinical radiology

3

Syamantak Mookherjee[a], Devjani Ghosh Shrestha[b], Skandesh Mohan[c]

[a]*Department of Radiology, Royal Liverpool and Broadgreen University Hospitals NHS Trust, Liverpool, UK*

[b]*Department of ENT and Head Neck Surgery, Bhagirathi Neotia Women and Childcare, Centre, Newtown, West Bengal, India*

[c]*Department of Radiology, Royal Derby and Burton University Hospitals NHS Foundation Trust, Derby, UK*

3.1 Introduction

Medical ultrasonography is an imaging technique developed by utilizing the properties of interaction of ultrasound with tissues and generating high-quality images for diagnostic purposes. The advances in ultrasound applications have been ever growing in the medical field since its inception.

3.2 Properties of sound

Sound is a form of mechanical energy that can propagate through a compressible/elastic medium in the form of a wave. Sound wave can potentially travel through any medium that has atoms and molecules; in fact, the speed of sound increases as the material gets denser (e.g., sound can travel well through air, but it travels faster through liquids and solids because of its relative increase in molecular density). Since propagation of sound requires atoms and molecules it cannot travel in vacuum. As the mechanical energy in the form of sound wave passes through a medium it vibrates the molecules and squeezes them into a tighter group known as the zone of compression. These zones alternate with areas where partial density is reduced due to the compression effect. These are known as zones of rarefaction. Hence, simply described, the sound is transmitted from one place to another by creating alternate zones of molecular compaction and sparsity. This physical effect of sound propagation can also be represented graphically in the form of a sinusoidal wave with its peaks and troughs corresponding to zones of compression and rarefaction, respectively (Fig. 3.1). The distance between two bands of compression or rarefaction is called wavelength. It is conventionally represented by the symbol λ. The number of cycles of compression and rarefaction repeats per unit time is called the frequency,

Biomedical Imaging Instrumentation. DOI: https://doi.org/10.1016/B978-0-323-85650-8.00011-5

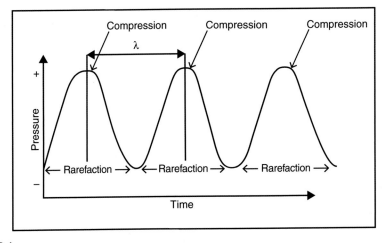

FIG. 3.1

Schematic representation of propagation of ultrasound wave with compression and rarefaction of medium it is passing through.

conventionally represented by the symbol μ. The greater the number of cycles per second, the higher the frequency and the higher the perceived pitch of that sound (within the audible range). The unit of acoustic frequency is hertz (Hz), where 1 Hz = 1 cycle/sec. Range of audible frequency is different for different animals (e.g., the human ear under perfect condition and physiologic state can hear frequency between 20 and 20,000 Hz; dogs, other mammals, and aquatic animals can perceive significantly higher frequency sounds). Ultrasound refers to all acoustic energy with a frequency above human hearing (20,000 Hz or 20 kilohertz [kHz]). In medical imaging the ultrasound transducer typically emits a range of frequency between 2 million Hz and 15 million Hz (2–15 megahertz [MHz]), far above the limits of human hearing.

The amplitude of a sound wave refers to its strength or level of pressure it exerts on molecules within the medium. It also corresponds to the number of molecules displaced by vibration created by the sound wave. It is the height of the sine wave that produced the sound. In audible sound, amplitude is analogous to loudness and is measured in decibels (dB).

The speed of sound varies depending on molecular density and physical characteristics (especially temperature) of the medium it travels through and is independent of frequency.

As the velocity of sound for a particular medium is constant under controlled conditions and is proportional to the density of the tissue, when frequency increases, the wavelength must decrease. The velocity of sound through a medium with low molecular density, such as gas (air), is much less than the speed through fluid. Most human tissue, except bone, behaves like a fluid medium with average velocity of 1540 m/sec. Sound (and ultrasound), like any other wave, has physical properties that govern its behavior within a given medium when it comes across an obstacle or when it encounters another medium of different molecular density. The ones that are most important from medical imaging and image formation perspectives are described in the next section.

3.3 **Interaction between ultrasound and matter**
3.3.1 **Reflection**

When ultrasound strikes the boundary between two media where there is a change in density or compressibility (or both), it gets reflected. The reflection occurs where there is a difference of acoustic impedance (Z) between the media (see Fig. 3.1). The impedance is a measure of how readily tissue particles move under the influence of the wave pressure and is a constant for a particular tissue.

Reflection is a fundamental property of any wave, and it is the most important and desired aspect of sound. Sound is reflected when it strikes the interface between two media of different particle densities or elasticities because of the difference in inherent susceptibility of tissue particles between the media to vibrate under the influence of propagating wave pressure, which in turn is responsible for the difference in velocities in each tissue or medium. In physics this is known as acoustic impedance, and it is of extreme importance for image formation and choosing the location for probe placement. For example, the different shades of gray that are projected on the ultrasound monitor are in fact produced due to subtle differences in acoustic impedance and resultant reflection of sound between those tissue interfaces and when everything taken together produces images of an organ in real time. However, if the acoustic impedance between the adjacent media (e.g., muscle and bone) is too great, then the ultrasound wave is almost entirely reflected, resulting in loss of tissue details beyond the interface. The greater the difference, the greater the percentage reflected. Difference between most body structures is fairly small except between air and bone. Reflection of sound waves off surfaces can lead to one of two phenomena: an echo or a reverberation. Echo is a desirable property of the reflection, which is peaked by the image transducer and formed image. Conversely, reverberation is not very useful in medical ultrasound except for very few specific instances and is generally considered an artifact.

3.3.2 **Refraction**

The change in direction of a beam when it crosses a boundary between two media in which the speeds of sound are different is called refraction (Fig. 3.2). If the angle of incidence is 90 degrees, there is no beam bending, but at all other angles there is a change in direction. Refraction is important because it leads to artifacts. If this is suspected, increasing the scan angle so that it is perpendicular to the interface minimizes the artifact.

Refraction can be easily calculated using snell's law, which states that the ratio of the sines of the angles of incidence and refraction is equivalent to the ratio

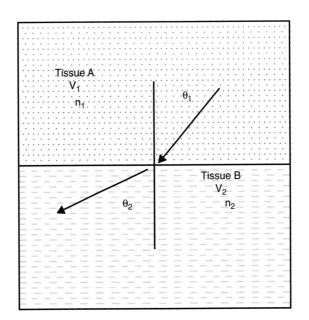

FIG. 3.2

Schematic representation of refraction of ultrasound wave between two media of different refractive indices (n_1 and n_2) with angle of incidence θ_1 and angle of refraction θ_2 and V_1 and V_2 are velocities of sound in respective media.

of velocities of sound in the two media, or equivalent to the reciprocal of the ratio of the indices of refraction:

$$\frac{\sin\theta_2}{\sin\theta_1} = \frac{v_2}{v_1} = \frac{n_1}{n_2}$$

where each θ is angle measured from the medium interface, v is velocity of sound in the respective medium and n is refractive index of the respective medium.

The results in ultrasonic imaging allow us to conclude that refraction is not a severe problem at soft tissue interfaces. An exception is at soft tissue/bone boundaries, where it can be a severe problem due to the big differences in speed of ultrasound between soft tissue and bone. When the sound wave reaches boundary between two media, part of it gets reflected as described in the previous section; however, the other part may pass into the next medium and continue traveling in it with a different speed because the speed of sound is dependent on elasticity and inertial properties of the medium as discussed in the previous section. The change in speed changes the wavelength (for a constant frequency) and the direction of wave. This change in direction of wave is called refraction. Since most of the soft tissue has similar composition, refraction is not a significant problem in medical ultrasonography except for creation of occasional artifact, which can be corrected by changing the transducer placement.

3.3.3 **Absorption**

Sound is an energy that propagates by vibrating the molecules, and in doing so it performs a work and gets converted to thermal energy according to laws of thermodynamics. The higher the frequency, the more quickly it vibrates molecules and gets dissipated as heat. Hence higher frequency transducers can only be used for superficial tissue, and for imaging deeper tissue a low-frequency transducer is needed.

Diffraction and interference. Diffraction is one of the fundamental wave behaviors and is equally applicable for electromagnetic and sound waves. By definition, diffraction is bending of waves around small obstacles and the spreading out of waves as the waves come out of the source or pass through a small opening, provided the obstacle is larger and the hole is smaller than the wavelength. Diffraction occurs because of the phenomenon called interference, which is constructive or destructive interaction of two or more waves that are created when the native wave hits an obstacle or moves forward through a hole. Diffraction is also encountered when ultrasound waves generated within the transducer pass forward, and hence are of great importance for shaping of the ultrasound beam.

Scattering. Scattering as a phenomenon is observed both for wave and particles. When a sound wave traveling through an inhomogeneous medium strikes a structure of different acoustic impedance to the surrounding medium, which is smaller than the wavelength of the incident sound wave, it gets reflected. This is known as scattering. Scattering can be minimized, but it is unavoidable and observed in all ultrasound images as background noise and loss of resolution. It is of most importance, however, for imaging of blood vessels where red blood cells act as a reflector causing scattering of waves.

Doppler shift. The Doppler shift or Doppler effect is defined as the change in frequency of sound wave due to relative motion between the receiver and source of sound. It is commonly perceived as a change of pitch, such as when a whistling train approaches and recedes from a bystander on the platform. When sound is emitted and subsequently reflected from a source that is not in motion, the frequency of the returning echo will equal the frequency at which it was discharged. However, if the reflector or the source of sound is in relative motion, the frequency of the sound waves received will be higher (positive Doppler shift) or lower (negative Doppler shift) than the original frequency depending on the relative speed between the emitter and receiver. In medical ultrasound, the receiver and transmitter are fixed, however, if the reflector is in motion (e.g., blood in arteries or veins will show a Doppler shift, which can be used to calculate the speed and direction of flow and is of fundamental importance for vascular imaging).

Doppler shift can be calculated by using the following formula for visual representation of data.

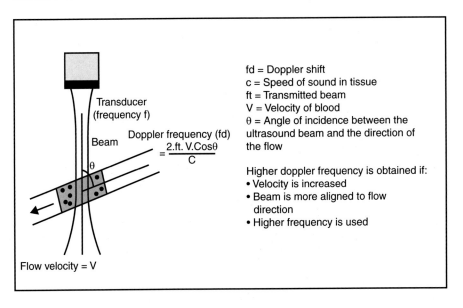

Transducer
(frequency f)

Beam

Doppler frequency (fd)

$$= \frac{2.ft.V.Cos\theta}{C}$$

Flow velocity = V

fd = Doppler shift
c = Speed of sound in tissue
ft = Transmitted beam
V = Velocity of blood
θ = Angle of incidence between the ultrasound beam and the direction of the flow

Higher doppler frequency is obtained if:
• Velocity is increased
• Beam is more aligned to flow direction
• Higher frequency is used

3.4 Fundamental technology

Fundamental to generation of ultrasound imaging is conversion of electrical energy to sound energy and reconversion of sound energy in the form of echo from soft tissue into kinetic or mechanical energy. These processes are known as piezoelectric and inverse piezoelectric effects.

3.4.1 Piezoelectricity and production of ultrasound

Piezoelectricity (derived from the Greek, "pressing electricity") is the electrical charge (electron) that gets accumulated within a material when it is subjected to mechanical stress, leading to generation of voltage that can be detected electronically. The piezoelectric effect is a reversible phenomenon, meaning the materials that exhibit the piezoelectric effect (voltage from stress) also exhibit the reverse piezoelectric effect (stress from voltage). The piezoelectric materials are often sensitive to high heat, and when heated above a certain temperature, called its Curie temperature, the transducer loses its piezoelectric properties. Therefore transducer probes are not sterilized by autoclaving.

Mechanism. The nature of the piezoelectric effect is closely related to the occurrence of electric dipole moments within the molecular lattice structure of certain solid materials. In an uncompressed relaxed state, ionic charges within those molecular lattice just cancel each other out, hence the material remains electrically neutral. However, when it is compressed in a certain orientation, the relative position of ionic charges within the molecule shifts in a way that positive charges accumulate in one end and negative charges in other, leading to creation of net vector charges on the opposing face of the lattice, which can be electrically measured as a voltage. The opposite is also true if electrical charge is applied on two opposing faces of the crystal; it repels electrons from one end and attracts from the other end, and vice versa, for positively charged ions, which ultimately leads to crystal deformation and mechanical stress, which can be released as ultrasound energy as well as many other forms of energy.

History. The piezoelectric phenomenon was first demonstrated in 1880 by French physicists Pierre Curie and Jacques Curie. The technology had its first application in the form of a sonar detector, and was used during World War I. Since then, piezoelectric effect has been used in many different products and industries, including medical ultrasound from the late 1940s that continues to evolve today.

Material. Piezoelectric materials can be both naturally and artificially made. Examples of some natural materials include various crystals (quartz, topaz, cane sugar, etc.). Artificial piezoelectric materials, called ferroelectrics, were first developed during World War II and include barium titanate (BTO) and lead zirconate titanate (PZT), among others, which are extensively used as piezoelectric discs for construction of ultrasound transducers.

Illustration showing Piezoelectric and reverse piezoelectric effect. First imaged demonstrates deformation of piezoelectric crystal upon application of voltage. 2nd image illustrates generation of electrical signal from piezoelectric crysta upon application of pressure by sound wave.

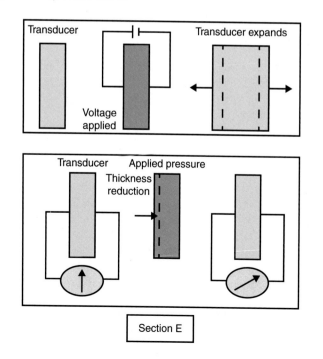

Section E

3.5 Instrumentation and image formation

The basic parts of a medical ultrasound include:

Transmitter
Beam former
Transducer
Receiver
Image processor/scan converter
Image display

3.5.1 Transmitter

The ultrasound transmitter is a specialized electronic circuitry that supplies the transducer with high voltage for the production of acoustic energy. It also controls the rate of electrical impulse applied on the transducer, which ultimately controls the pulse repetition frequency (PRF) of the ultrasound beam, which determines depth of tissue that can be imaged without artifact.

3.5.2 Beam former

The beam former is responsible for sending electrical voltage from the transmitter to the transducer piezoelectric elements in ways that the resultant ultrasound beam could be controlled, steered, and focused in a specific angular direction.

3.5.3 Transducer

Broadly speaking, a transducer is a device that converts one form of energy to another. An ultrasound transducer holds an electromechanical assembly that converts electrical energy received from the transmitter and beam former into mechanical (sound) energy and sends it to the body. It also converts the weak mechanical energy in the form of returning echoes from the body back into electrical energy.

3.5.3.1 *Parts of the transducer*

Piezoelectric crystal (PTZ). The most important component of a transducer is the piezoelectric crystal, which is described in an earlier section.

Two electrodes. The back and front of the crystal are in contact with thin films of electrically conducting material to facilitate supply of voltage necessary for pulsing the piezoelectric crystal. In addition, the piezoelectric signal generated by returning echoes from a patient's body strikes the crystal. Electrodes also serve to pick up the piezoelectric signal generated when returning echoes strike the crystal. The front side of the transducer, which makes direct contact with the patient, is covered with an electrical insulator.

Backing Block. The backing block is placed behind the crystal layer and positive electrode for the purpose of absorbing the ultrasound energy directed backward during vibration of PTZ and attenuates unwanted ultrasound signals from vibration of housing.

Matching Layer. The matching layer is placed in front of piezoelectric ceramic elements to provide the required acoustic impedance gradient for the acoustic energy from the transducer to smoothly enter the body and for the returning echo to enter the transducer. The main purpose of the matching layer is to reduce large impedance difference between the acoustic source and target, which reduces the loss of transmission and receipt of acoustic energy.

Acoustic insulator. This component of rubber or cork prevents the sound from passing into the housing. The acoustic insulator prevents vibrations originating from the crystal from being transmitted into the transducer housing.

Housing. The internal electrical and crystal components of the transducer are covered by a robust housing usually made of strong plastic.

3.5.3.2 *Probe construction*
Probe construction with a schematic image is shown in Fig. 3.3.

3.5.4 **Receiver**

An ultrasound receiver picks up the weak electrical signal from the transducer and preprocesses it for further processing through multiple steps, which includes amplification, compensation, compression, demodulation, and rejection. Finally, the preprocessed signal is sent off to the image processor.

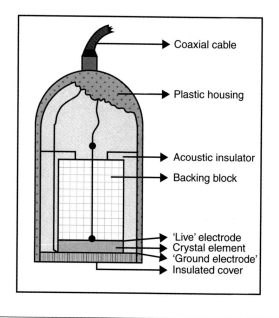

FIG. 3.3

Probe construction.

3.5.5 Image processor/scan converter

The receiver sends the preprocessed images to the image processor or more conventionally to the scan converter, which further processes the image and sends it to the image display unit.

3.5.6 Image display

Traditionally the image was displayed on a cathode ray tube (CRT) monitor. With the advent of newer flat-panel monitors, most current machines use a high-resolution liquid crystal display (LCD) or light-emitting diode (LED) monitor for image display.

3.6 Image representation and clinical application

Ultrasound image is a visual representation of data generated by scanning a volume of tissue. Like any data it can be represented by various ways to highlight certain properties of the imaged tissue. Over the years through the gradual evolution from the A-mode and M-mode, other advanced modes of displays were developed, which are described briefly in this section.

3.6.1 Modes of ultrasound

3.6.1.1 A-mode (amplitude mode)

A-mode is the simplest and earliest form of representation of ultrasound data.

In the A-mode representation, the signals are displayed as vertical deflections on horizontal temporal base tracings. The amplitude of the deflection is proportional to the strength of the returning echo. Thus a tissue interface composed of substances of high acoustic impedance, such as a tissue-bone interface, will produce almost total reflection and the strongest echo; a solid-liquid interface will produce a relatively weak echo. Distance between the echoes may be used to calculate the depth of origin of echo based on knowledge of the velocity of ultrasound in tissues. This mode of data representation has largely been replaced with current methods.

Clinical Application: A-mode in ultrasound is of historical significance and is no longer commonly used.

3.6.1.2 M-mode (motion mode)

The M-mode is a form of representation that captures the motion of structure in the scan path. All the reflecting interfaces are displayed along the time axis, whereas the brightness of each dot is proportional to the amplitude of reflected waves. It is still in extensive use.

Clinical Application: M-mode has use in adult cardiac imaging, estimation of fetal heartbeat, measurement of inferior vena cava diameter and collapsibility, diaphragmatic excursion with respiration, among others. It also has use in pleural imaging.

3.6.1.3 *B-mode (brightness mode)*

In B-mode, the returning echoes are displayed as gray scale value of varying intensity instead of amplitude (as in A-mode). When multiple such values are created by an array of PTZ crystals along the transducer face and displayed next to each other, an image is formed. The brightness of the displayed image is proportional to the strength of returning echo, while the less reflective areas appear darker.

Real-time high-resolution rapid B-mode scanning is the most frequently used mode in today's world. Multiple piezoelectric crystals work in concert to create a moving two-dimensional (2D) image.

An array of crystals produces beams along the scan plane, each one producing a static B-mode image. Like moving pictures or cinematography, when these static B-mode images are acquired at rapid succession or 25 frames/sec or beyond, motion of tissues or real-time experience is perceived by the operator.

Instead of planner 2D images, volume data of a tissue can be acquired by a specialized volume probe where the PTZ is arranged in a rectangular volume axis. The acquired dataset may be viewed by reformatting later (called 3D mode) or may be viewed in real time (called 4D mode).

B-mode (2D live and freeze) is used for real-time scanning of tissue.

During the scanning, brightness and contrast of the image can be adjusted by changing gain setting.

Live B-mode scanning is the workhorse of most scanning for all types of probes. It is used to make mental impression by the operator. At any point of time the image on the screen can be frozen and captured (saved) for viewing later in electronic format. Images can also be printed directly if attached to a specialized thermal printer.

This mode is used for basic assessment of all organs and soft tissues.

Contrast enhanced ultrasound is a specialized application of B-mode in which specialized microbubble contrast is injected through the vein while real-time scanning of a suspected lesion is carried out (e.g., a liver lesion) for better characterization of the lesion.

Clinical Application: Real-time 2D B-mode imaging is the most widely used form of ultrasound. It is used in abdominal and pelvic imaging, fetal imaging, imaging of soft tissue and musculoskeletal system, intracavitary, intravascular imaging with a wide range of applications through various kinds of curvilinear, linear, and other specialized transducers.

3.6.1.4 *Doppler ultrasound*

In Doppler mode, the Doppler frequency shift of moving objects (described earlier) is used for image formation in the form of color or spectrum. There are two different types of Doppler modes: pulsed wave and continuous for specific clinical application. Using these two types of mode-collected data can be represented as spectral wave form or color. Another specialized form of pulsed Doppler is tissue Doppler, which is highly sensitive for detection of vascularity within the tissue. The Doppler signal data are usually superimposed on real-time gray scale images or represented along with a frozen section of the gray scale image.

Clinical Application: Doppler ultrasound has extensive application in vascular imaging of limb vessels, carotids, and aorta, and for assessment of fetal vessels and fetal-maternal circulation. Echocardiography is heavily reliant of Doppler ultrasound.

3.6.2 Elastography

Ultrasound elastography is a specialized imaging technique that detects tissue stiffness and elasticity and presents it to the viewer as a color map or in the form of measurement. Based on this data, hardness of a lesion can be assessed, which aids in diagnosis and helps decide the best site for ultrasound-guided biopsy. This technology has extensive use in assessment of liver and breast nodules. Musculoskeletal elastography is being used with some success.

3.7 Conclusion

Medical ultrasound has immense importance as a diagnostic and therapeutic modality in clinical medicine. Its clinical applications are constantly evolving and expanding. Artificial intelligence–based ultrasound is now a reality and promises to transform our health care system in numerous ways. An understanding of the basic principles of ultrasound is a prerequisite to developing the ability to interpret ultrasound images accurately. It will enable the sonographer to maximize the potential of this complex yet versatile modality.

Magnetic resonance imaging: Basic principles and advancement in clinical and diagnostics approaches in health care

Doniparthi Pradeep[a,1], Manoj Kumar Tembhre[b], Anita Singh Parihar[c], Chandrabhan Rao[d]

aDepartment of Cardio-Thoracic & Vascular Surgery, AIIMS, New Delhi, India
bDepartment of Cardiac Biochemistry, AIIMS, New Delhi, India
cDepartment of Dermatology & Venereology, AIIMS, New Delhi, India
dAbhinn Innovation Private Limited, Jaipur, Rajasthan, India
[1]Note: a. First three authors have contributed equally; b. Corresponding authorship will be shared between Dr. Chandrabhan Rao and Dr. Manoj Kumar Tembhre

4.1 Introduction

The concept of magnetic resonance and its implication in imaging was first introduced by Paul C. Lauterbur in 1973 who proposed the dimensional/spatial knowledge of the object using the basic principles of zeugmatography (image formation technique, derived from Greek word ζεῡγμα meaning "that which is used for joining") and proton nuclear magnetic resonance (NMR) to develop the zeugmatograph (relative orientation of the object along the gradient directions) of the object (Lauterbur, 1973). The NMR technique was developed in 1946 by Edward M. Bloch and Felix Purcell, and both were awarded the Nobel Prize for Physics in 1952 for their discovery (Bloch et al., 1946; Purcell & Torrey, 1946).

The technique of magnetic resonance and image acquisition was later polished by Peter Mansfield in the 1970s. His team developed images of the body using NMR by inventing "slice selection" for MR-based imaging (explained the concept of conversion of radio signals using mathematical algorithms to produce meaningful images and its interpretation). For their remarkable contributions in the development of MRI, the duo of Paul C. Lauterbur and Peter Mansfield were awarded the Nobel Prize in Physiology or Medicine in 2003 (Slavkovsky & Uhliar, 2004). The clinical MRI scanner was first introduced in the health care practice in the early 1980s, and thereafter rapid technological development and advancement happened that has revolutionized as a routine test in the field of medical diagnostics. Due to its immense imaging potential and superiority to other imaging techniques, today MRI has become an integral part of almost all health care setups. Every year more than 60 million investigations are performed by MRI to obtain significantly improved diagnostics in many diseases. To reduce the risk and discomfort of many patients, MRI serves as a powerful noninvasive diagnostic imaging tool by replacing various invasive examinations, and it has gained wide attention in the era of modern medicine as it is capable of efficiently producing cross-sectional images of complex anatomic structures (Hawkes et al., 1980; Smith et al., 1981; Slavkovsky & Uhliar, 2004).

MRI technology exploits the basic principle of NMR that uses nonionizing electromagnetic radiation, where protons (atomic/hydrogen nuclei) present in the object/specimen absorb and reemit the electromagnetic waves under the influence of a strong magnetic field. The hydrogen nucleus is found in water (H^+—O^-—H^+) and therefore such nuclei are found in all types of tissues of the body enabling the application of the concept of magnetic resonance for imaging of any body parts. The hydrogen nuclei act as a magnetized pointer (similar to a compass needle oriented in north-south direction under earth's magnetic field) that is aligned in characteristic spatial orientation under the effect of the strong magnet of the MRI scanner. The radiowaves rotate the hydrogen nuclei making them oscillate in the magnetic field when the magnetic field returns to an equilibrium state. The emitted electromagnetic waves of a particular frequency of the magnitude of radiofrequency (40–130 MHz) are received by receiving coils, and radiofrequency signals were then converted into positional details of the object by interpreting the changes in radiofrequency level and phases (via gradient coils) (Westbrook et al., 2011; Grover et al., 2015). MRI is considered safe compared to other imaging tools as it does not involve the use of radioactive isotopes and ionizing radiation, and there are no adverse side effects. MRI is a highly flexible technology that enables the measure of very fine details of soft tissues, oblique orientation imaging with two-dimensional (2D) and three-dimensional (3D) orientations, and aids in revealing the structural and functional information. Due to its wide clinical applications MRI tool usage has been extended to multiple disciplines of medicines such as cardiovascular, neurology, oncology, gastroenterology, and musculoskeletal structures (Galbán et al., 2017; Meijer et al., 2017; Otero-García et al., 2019; Sotoudeh et al., 2016). This chapter will provide an insight into the basic principles of MRI, as well as components of the equipment, applications, and advancements in the technology, limitations, and future prospects.

4.2 Basic principles of MRI

Understanding the basic principles of MRI requires thorough knowledge of the quantum physics of an atom. As mentioned, the MRI uses the fundamental basis of NMR technology; it can be explained by knowing the properties of atomic nuclei and the constituents. Atomic nuclei (consisting of protons and neutrons) have the tendency to spin as if the nucleus is spinning around its own axis. The nuclei cannot spin on their own but do so under the influence of internal charge components. The atomic nuclei behave as bar magnets having a local north-south pole inducing a magnetic dipole moment (Fig. 4.1) (Westbrook et al., 2011).

Under the influence of a strong external magnetic field (B_0) the nucleus is aligned in a perpendicular or parallel direction of B_0. Depending on the nuclei constituents of the specimen there may be multiple nuclear spins within a particular B_0 at varying energy states (i.e., high-energy state aligned perpendicular to B_0 and low-energy state aligned parallel to B_0). In contrast to a bar magnet that is aligned either parallel or antiparallel B_0, atomic nuclei possess angular momentum and will orient around the B_0 axis. Such spin is comparable to the wobble (**precession**) motion of a gyroscope under the influence of the magnetic field of the earth (earth magnetic field = 0.5 Gauss or 0.00005 tesla [T]). This phenomenon can be mathematically expressed by the Larmor equation, where the velocity of spin around the applied field (referred to as Larmor [**precession**] frequency) is proportional to the strength of the applied field (Grover et al., 2015; Westbrook et al., 2011).

$$\omega_0 = \gamma B_0$$

where ω_0 = Larmor frequency (MHz), γ = gyromagnetic ratio in (MHz/T), and B_0 = magnetic field strength (T).

In MRI, the **Larmor (precessional) frequency** is defined as the velocity of the precession of the magnetic moment of atomic nuclei (protons) around the applied external B_0. The gyromagnetic ratio is a constant that is fixed for a specific nucleus (e.g., γ for commonly used isotopes are H^1 = 42.58; F^{19} = 40.05; Na^{22} = 11.26; and P^{31} = 17.24).

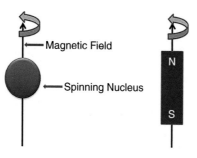

FIG. 4.1

Schematic representation of spinning nucleus *(left)* and induced magnetic dipole (with N = north, S = south alignment) where the nucleus behaves like a bar magnet as a result of spin *(right)*.

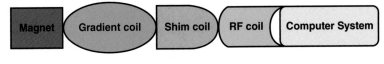

FIG. 4.2

Schematic representation of the basic hardware components of the MRI system.

4.3 Components of MRI

The MRI system is composed of five basic components: magnet, gradient coils, shim coils, radiofrequency coils (transmitter and receiver), and the computer system as shown schematically in Fig. 4.2. These components work in concert to generate high-contrast cross-sectional imaging of the specimen. The overview of a clinical MRI system is shown in Fig. 4.3.

4.3.1 Magnet

The magnet is regarded as the heart of the MRI system and it is the most expensive component of the scanner. There are three different types of magnets used in the MRI systems:

1. Resistive magnets
2. Permanent magnets
3. Superconducting electromagnets

Resistive magnets possess an inherent resistance similar to any electrically conducting material, and such magnets were made of copper. However, when electric current flows through such resistive magnets to produce a strong magnetic field, it generates a considerable amount of heat that is the major limitation of these types of magnets and can be used for low field strength applications. Unlike resistive magnets that produce magnetic fields upon passage of electric current, the permanent magnet does not require an electrical supply for the generation of magnetic fields, and coolants are also not required for functioning. Like resistive magnets, the permanent magnet also suffers from the disadvantage of low field strength. However, both types of magnets offer more patient access and are suitable for generating a vertical magnetic field. Resistive electromagnets and permanent were used in the conventional MRI system, but superconducting magnets are widely used in modern systems as they offer high field strength for better image resolution. The resistive and permanent magnets for whole-body imaging offer field strength of 0.3 T, and it is considered ideal for low field applications providing easy patient access. The superconducting magnet offers field strength of about 10 to 1000 times in magnitude compared to the resistive electromagnets and permanent magnets, and a scanner of 4.0 T is now commercially available for clinical applications. The material used to construct a superconducting magnet are generally niobium-titanium (Nb-Ti) alloy sandwiched between copper wires (act as insulation) that loses its resistivity at a temperature

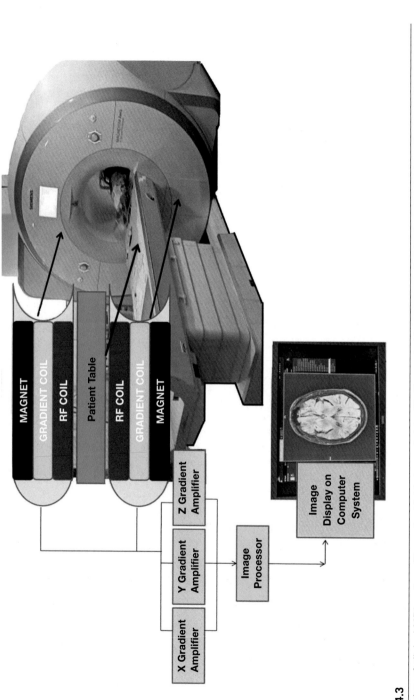

FIG. 4.3

Overview of clinical MRI system. The patient table is sandwiched in the central bore surrounded by the solenoidal arc that comprises radio-frequency (*RF*) coils, gradient (and shim) coils, and superconducting magnet. After magnetic induction, the signals were transmitted through various gradient amplifiers, and signals are decoded into three-dimensional images by the processer. Image is eventually displayed in the computer system and clinically correlated.

of around 9 K—a superconducting state (1 K = 273°C). The coolant used for such superconducting magnets for maintaining a specific temperature is liquid helium enabling the constant flow of current in the coils, and field strength generated is sustained and stable for a longer duration. Superconducting magnets are generally cylindrical or solenoidal in shape with a central bore region. To generate a strong and stable magnetic field, the bore diameter should be small, and a long bore size is required. This feature is a major limitation of superconducting magnets often causing patient discomfort (claustrophobia). The selection of the type of magnet depends on magnetic field strength, homogeneity, access to patients, and maintenance cost (Di Costanzo et al., 2003; Soher et al., 2007; Westbrook et al., 2011).

The magnetic field strength is expressed in terms of tesla (T) or gauss (G) (1 T = 10,000 G).

Today the MRI system used in clinical setup uses magnets offering field strength of 1.5 or 3.0 T but ultrahigh field strength of 7 T (even 10.5 or 11.7 T) have also become popular in clinical research purposes. However, sensitivity and optimizing signal-to-noise issues is one of the disadvantages of such systems. The configuration of the magnet used in the MRI system uses a solenoidal model in geometry where the patient/specimen is placed in the center, which ensures uniform and strong magnetic field minimizing the shift in Larmor frequency (Di Costanzo et al., 2003; Soher et al., 2007).

4.3.2 Gradient and shim coils

As mentioned, homogeneity of the applied magnetic field is critical in MRI and is ensured by the use of shim coils, a phenomenon called shimming, which simply refers to the adjustment of field strength to maintain uniformity. The shim coil compensates the magnetic field and minimizes the spatial vibration. The linear variations in the magnetic field are adjusted by the gradient coils, which also help in translating the spatial information of the specimen into signals. To retrieve the spatial configurational information of the object or any specific organ of the body along the arbitrary x-y-z orthogonal axis, three gradient (Gx-Gy-Gz) coils were used in combinations. Gradients were created by passing the electric current through the gradient coils. The magnitude of the MRI signal depends on the precession speed. To cover a large volume across the specimen in less time, high slew rate and high gradient were used for short repetition time and echo time.

4.3.3 Radiofrequency coils (RFC)

The MRI system is equipped with RFC (transmitter and receiver) that act as antennae of the broadcasting station that transmits the signals to the patient's body or area of interest and receives the signal coming out of the target tissue/organ. There are various types of RFC used in clinical MRI systems (e.g., surface coil, gradient coil, paired saddle coil, Helmholtz pair coils [two parallel coils, also used as gradient coils], birdcage coil [offers the best homogeneity compared to all RFC, act as both transmitter and receiver coil], transverse electromagnetic [TEM] coil [Fig. 4.4] that

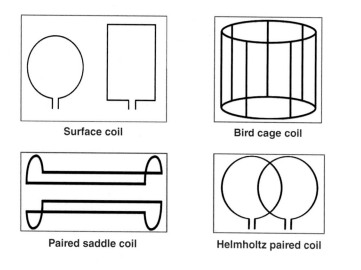

FIG. 4.4

Schematic representation of commonly used radiofrequency coils (RFC) in the MRI system.

can be broadly categorized as surface and volume coils [the former is suitable for the small surface area of interest and the latter is used for the larger surface area]) (Avdievich, 2011; Corea et al., 2016; Junge, 2012; Keil et al., 2013; Nordmeyer-Massner et al., 2012; Requardt et al., 1987; Wiggins et al., 2016; Wright et al., 2011). These coils are used depending on the type of specimen (tissue/organ) to be analyzed (e.g., surface coils [single or array coils] preferred for the whole body and body parts such as head, neck, chest, breast, abdomen, spine, shoulder, temporomandibular, rectum, prostate); the birdcage coils are suitable for head neck and knees; Helmholtz pair coils are used for pelvis and cervical region. The RFC are critical components of the MRI system as they largely determine the homogeneity of the magnetic field, sensitivity, and resolution of the image.

4.3.4 Computer system

The MRI scanner is interfaced to the computer system, which facilitates the simulation of the acquired signal into the reconstruction of a high contrast image that is eventually displayed on the screen.

4.4 The basic concept of image acquisition and formation in an MRI system

Before understanding the process of image acquisition, the knowledge of some basic terminologies is necessary. As explained earlier, image quality is directly proportional to the strength of the applied external field, and it offers the advantage of enhanced signal-to-noise ratio and higher resolution (in terms of spectral and spatiotemporal aspects) but suffers from the disadvantage of magnetic field instability and artifacts produced by eddy current (Foucault currents generate as a result of a change in the magnetic field and flow in closed circular loops inside the conductor). Magnetic susceptibility, which is the measure of magnetization capacity of tissue, body fluid, or objects under the influence of external magnetic field, is also a limitation accompanied by larger field strength ($\geq$3 T), and these factors largely determine image quality. Depending on the excitation of atomic nuclei present in a target tissue slice or slab, 2D or 3D images are acquired. After excitation of nuclei, target tissue emits signals along with the slice/slab in the form of phases or frequency, which is decoded by magnetic gradients (coils). Such selective excitation is achieved via the application of RF pulse, and spins having frequency equivalent to Larmor frequency will be selectively excited. The magnitude of amplitude and bandwidth of the RF pulse depends on the thickness of the tissue. The spatial information of spins that are hidden in the frequency signals and phases are decoded by applying the magnetic field gradient along the direction of the magnetic gradient (during signal acquisition) or in the form of a short pulse (before signal acquisition), respectively. The position information is retrieved from the phases, and the abovementioned process of signal encoding is repeated multiple times by applying the gradient pulses of varying amplitudes. To generate the 2D or 3D image, different combinations of frequency and phase decoding are employed, and the target tissue slice is repeatedly excited. In doing so, the amplitude of the phase gradient is increased with each repetition while keeping constant the amplitude of the frequency gradient. The resulting spectral lines arranged in 2D or 3D arrays (termed k space, where k = wave number or vector, its orientation revealed the direction of movement of nuclei) are recorded as output signal, and spatial information is deciphered using 2D or 3D Fourier transform (a mathematical transform used to define the constituent frequencies of a function of signal/time) (Fig. 4.5). The k-space spectral data are critical in determining the image quality; the spectral data lying near to outer edges (having high spatial frequency) correspond to fine structural details of the object, whereas spectral data close to the center of the k space (possess low spatial frequency) correspond to coarse/large structural details.

Radiofrequency (ω_{RF}) at Larmor precessional frequency (ω_0) of the hydrogen (atomic nuclei) applied perpendicular to B_0 (external magnetic field) resonates and flips the net magnetization vector into transverse plane. The individual magnetic dipoles of hydrogen nuclei aligned in phase and coherently process at ω_0 in the transverse plane. The signal (having frequency equal to ω_0) is induced in the receiving coil, which is then spatially located in three dimensions (first locating the slice or slab, which is then encoded into both the axes of image) via magnetic gradients followed by a combination of frequency and phase encoding (i.e., locating the MR signal as a function of time and positioning it correctly on image). The recorded MR signals are represented as an array of spectral lines (k space). By applying Fourier transform the image is constructed. *k*, Vector (constant); *M*, net magnetization

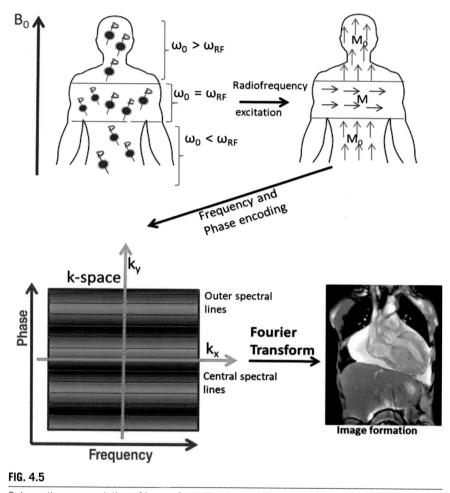

FIG. 4.5

Schematic representation of image formation in the MRI imaging system.

(at equilibrium it is aligned along with magnetic field B_0); M_0, magnetization at equilibrium (aligned perpendicular to magnetic field B_0).

Another important aspect of MRI is the image resolution that is determined by the recovery of the MR signal, and it depends on relaxation time. The relaxation time is defined as the time required for spinning nuclei to return to the equilibrium after RF excitation. Two types of relaxation time constants (i.e., T_1 [longitudinal or spin-lattice relaxations] and T_2 [transverse or spin-spin relaxations]) are generally used in MRI. This T_1 refers to the duration of time the magnetization returns to equilibrium in the longitudinal direction; it differs with tissue type (e.g., longer T_1-weighted image [T_1WI] indicates hypointense tissue). The T_2 refers to the duration required for a signal to decay after excitation in the transverse direction (e.g., longer T_2-weighted image [T_2WI] indicates hyperintense tissue). The T_1 relaxation time is always longer than or equal to T_2, as transverse magnetization decays faster than longitudinal relaxation recovery. In in vivo imaging, T_1 usually ranges from 300 to 3000 ms, whereas T_2 ranges from 10 to 200 ms; this range applies to most tissue types. However, longer T_2 are observed for water or body fluids (e.g., T_1 and T_2 water/cerebrospinal fluid are 4000 ms and 2000 ms, respectively). No signal intensity (appears black in MR image) is observed for cortical bones, calcified tissue, enamel, dentin, air, and blood vessels.

4.5 Contrast agents used in MRI

Contrast agents or materials are the group substances used in conjugation with various imaging techniques such as x-ray, computed tomography (CT), and MRI. The discovery of contrast agents has revolutionized clinical imaging and is frequently used to discriminate between the normal and abnormal states of tissue. In the early 1980s, after the introduction of clinical MRI, ferric chloride was first used as a contrast agent to produce contrast-enhanced human MRI, and it was used for the gastrointestinal tract (Young et al., 1981). In 1984, Carr et al. introduced gadolinium as a contrast agent for the first time that was administered intravascularly. Working knowledge of the magnetic properties of the substances is necessary before understanding the rationale of the MRI contrast agents.

4.5.1 Magnetic susceptibility

As explained earlier, all substances possess atoms having dipole moments, and it is the basic property of the substance to undergo magnetization when influenced by an external magnetic field. Based on magnetic properties, substances are grossly divided into four types:

1. **Diamagnetic substances:** These substances have negative susceptibility to the external magnetic field. These substances retain the external magnetic field and don't retain the magnetization when the external magnetic field is removed. Owing to the presence of paired electrons, these substances exhibit this property (gold, silver, copper, etc.).
2. **Paramagnetic substances:** These substances have slight positive susceptibility and show a positive effect on the local external magnetic field. These substances are attracted by the external magnetic field and do not retain magnetization when the external magnetic field is removed. This property is due to the presence of the unpaired electrons, which can undergo realignment (e.g., gadolinium, manganese).
3. **Superparamagnetic substances:** These substances have greater positive susceptibility than paramagnetic substances to the external magnetic field and render a greater degree of disruptive changes in the applied external magnetic field (e.g., iron oxides).

4. **Ferromagnetic substances:** These substances have high positive susceptibility to the external magnetic field and retain the magnetization even after the removal of the external magnetic field. The presence of magnetic domains in their atoms is responsible for this property (e.g., iron, nickel, cobalt). Paramagnetic and superparamagnetic substances are an area of interest in the clinical utility of MRI contrast agents.

4.5.2 Molecular tumbling/Larmor frequency and dipole interactions: A rationale of contrast

As explained earlier, an atom spins around its own axis, and the application of the external magnetic field makes the atom wobble and produces additional spin called secondary spin. The secondary spin is called precession and is responsible for the magnetic moments to follow the circular path around the magnetic field. The speed at which atoms wobble around the magnetic field is called precessional frequency or Larmor frequency. In other words, the entire molecule tumbles in the magnetic field, and at near Larmor frequency, the T_1 relaxation is short. Water molecules tumble much faster than Larmor frequency, hence they have longer T_1 relaxation times, which appear dark on T_1-weighted images. If a tumbling molecule with a more magnetic moment is inserted in the presence of water spins, a disruption in the local magnetic field occurs. Agents like gadolinium compounds create these fluctuations nearer to Larmor frequency and decrease the T_1 relaxation times of surrounding spins rendering the tissue/object brighter on T_1-weighted images.

4.5.3 Nature of contrast agents

T_1 **agents:** These substances decrease the T_1 relaxation time, so they appear much brighter in T_1- weighted images. These are positive contrast agents (e.g., gadolinium, manganese derivatives).

T_2 **agents:** These substances decrease the T_2 relaxation time, so they appear much darker in T_2- weighted images. These are negative contrast agents (e.g., iron oxide compounds).

4.5.4 Classification of contrast agents

Based on the route of administration, contrast agents are grossly divided into three types:

1. Intravenous
2. Oral
3. Inhalational

4.6 Intravenous agents

Intravenous MRI agents are again divided into three types: (1) extracellular fluid agents, (2) blood pool agents, and (3) target/organ-specific agents.

1. **Extracellular agents:** These are the most common contrast agents used in clinical practice. Owing to low molecular weight, these substances seep out from the vascular compartment into extracellular space. Extracellular agents are excreted by the kidney. These agents are used to interrogate the endothelial lining of the structures such as disrupted blood-brain barriers (e.g., gadoterate meglumine).
2. **Blood pool agents:** These agents are restricted to the intravascular compartment. This is made possible by any of the following strategies:
   - Binding them with serum albumin, which is restricted to the intravascular compartment (e.g., gadofosveset trisodium).
   - Increasing the size of the particle to reduce the extravasation into the extra-vascular compartment (e.g., gadomer).
   - Increasing the intravascular circulatory time (e.g., small iron oxide nanoparticles such as ferumoxytol).
3. **Organ-specific agents:** Altering the elimination pathway and including the target organ of interest is the rationale of organ-specific contrast agents. These agents can target specific tissues such as the liver, spleen, and lymph node. Some commonly used organ-specific contrast agents are as follows:
   a. **Gadolinium derivatives** (gadoxetic acid, gadobenate for liver and biliary system)
   b. **Manganese(II) derivatives** (mangafodipir for imaging of liver, pancreas, and myocardium)
   c. **Iron oxide particles** (ferumoxide, ferucarbotran for the imaging of the reticuloendothelial system)

Oral contrast agents

These agents aid in the investigation of the gastrointestinal tract. Derivatives of gadolinium, manganese, barium sulfate suspension, and blueberry pineapple juice (abundant in manganese ions) are generally used in clinical practice.

Inhalational (ventilation) agents

Ventilation contrast agents are used to assess the ventilation in the lungs. Paramagnetic substances such as gadolinium-based aerosols and oxygen, hyper-polarized gases such as helium (^{3}He), xenon (^{129}Xe), and inert perfluorinated gases such as sulfur hexafluoride (SF_6) are some of the commonly used substances. Some of the contrast agents are administered by enema (i.e., via rectum). After the clinical imaging procedure is completed, most of the contrast agents are either absorbed by the body or eliminated externally via bowel movements or urination.

4.7 MRI contrast agents: based on the specific area of interest

4.7.1 Brain

Usually, structures outside the blood-brain barrier (BBB), extraaxial, such as falx cerebi, choroid plexus, pineal gland, pituitary gland, pituitary stalk, slow-flowing blood vessels, and sinus mucosa, show normal enhancement. Gadolinium aids in the diagnosis of extraaxial lesions such as acoustic neuroma, meningioma, and pituitary adenoma. Intraaxial lesions such as tumors of brain parenchyma and infarcts enhance due to breach in the BBB. Brain metastatic lesions can be identified with gadolinium.

4.7.2 Spine

Although spinal cord lesions can be identified without contrast, gadolinium enhancement gives better delineation. Spinal cord metastasis, syringomyelia, and multiple sclerosis are better diagnosed with gadolinium enhancement. Gadolinium aids in the differentiation of scar tissue from the reherniation in postdiscectomy patients who are symptomatic.

4.7.3 Abdomen

Gadolinium aids in the perfusion studies of the kidney, liver, spleen, adrenals, vascular structures, and pelvic structures. Although the liver gets the majority of blood supply from the portal vein, liver lesions depend on the hepatic artery. In the arterial phase liver lesions enhance first, in the portal venous phase normal liver parenchyma also enhances so they appear isointense.

4.7.4 Blood vessels

Gadolinium administration helps in the assessment of the blood vessels of the body. Arterial flow in the abdominal vessels is demonstrated with a contrast-enhanced MR angiogram.

Myocardial perfusion is demonstrated with dynamic perfusion sequences, which require gadolinium enhancement.

4.7.5 Breast imaging

Gadolinium enhancement is useful in the evaluation of breast malignancy and breast implant assessment. It can delineate the recurrence from scar tissue.

4.6.6 Safety concerns associated with the use of contrast agents

Although MRI is safe, there are some safety concerns on gadolinium contrast. A well-known, rare complication such as nephrogenic systemic fibrosis was noticed in a few patients (nearly 6%). It was noticed that patients with renal dysfunction are more prone to suffer from these complications between 5 and 75 days of exposure. The half-life of gadolinium (90 min) is increased in patients with renal impairment (18–34 hr). Glomerular filtration rate (GFR) of less than 30 mL/min/1.73 m^2 is a warning for the use of gadolinium in patients of renal dysfunction who are prone to nephrogenic systemic sclerosis. Although no detrimental effects are manifested, deposition of the gadolinium may be observed in the brain parenchyma, especially in the dentate nucleus in patients of multiple sclerosis who undergo repeated MRI.

4.8 Types of MRI and their applications

4.8.1 Anatomic imaging

Body fluids such as cerebrospinal fluid (CSF) have longer T_2 relaxation time than grey matter, which has longer T_2 relaxation time than white matter (i.e., CSF > grey matter > white matter). So ventricles (CSF) appear brighter in T_2-weighted images and darker in T_1-weighted images. T_2-weighted images aid in the diagnosis of brain tumors, which appear hyperintense. MRI is useful in the diagnosis of secondary parkinsonism resulting from multiple sclerosis, vascular lesions, and neoplasms. The application of the proper protocol that best suits the clinical scenario depends on the clinical history and indication for the examination. Fluid-attenuation inversion recovery (FLAIR) is a helpful pulse sequence that gives T_2-weighted images, in which usual high-intensity signals of CSF are suppressed. FLAIR images are more sensitive than conventional spin-echo images for lesions containing water or edema. To detect structural abnormalities and pathologic conditions in the brain by generating superb image resolution/contrast without radiation exposure, MRI and magnetic resonance spectroscopy (MRS) along with CT and positron emission tomography (PET) are used by clinicians (Liu, 2015).

4.8.2 Functional magnetic resonance imaging (fMRI)

fMRI is a blood oxygen level–dependent (BOLD) imaging modality based on acquiring ultrafast information about the blood flow and oxygen concentration. The fMRI is based on the principle of "the paramagnetic nature of the deoxyhemoglobin which diphase[s] the protons in the surrounding water." When the brain is active, increased blood flow leads to a rise in oxygenation, which in turn decreases the deoxyhemoglobin and increases the T_2 relaxation time. BOLD images can delineate the hyperactive foci from the surrounding normal brain parenchyma. fMRI is applied for the study of brain functions in healthy persons and in neuropsychiatric disorders (e.g., attention deficit hyperactivity disorder [ADHD], depression, obsessive-compulsive disorder [OCD]) (Labbé Atenas et al., 2018; Mitterschiffthaler et al., 2006). In the early disease course, to differentiate Parkinson disease and atypical parkinsonism is challenging due to its common symptoms. So, the brain MRI is applicable to assess cerebrovascular damage to improve the diagnosis with more accuracy and confidentiality. Simultaneously, brain MRI helps in finding the possible or probable diagnosis of a specific form of atypical parkinsonism (Meijer et al., 2017).

4.8.3 Diffusion tensor imaging

Diffuse tensor imaging (DTI) is an indirect quantification of the degree of anisotropy and orientation of structure that assesses the direction of microscopic motion of water along white matter tracts. Diffusion of the water molecules in white matter is limited along the axon, thus direction-dependent or anisotropic, whereas in the grey matter it is less anisotropic, and in CSF, diffusion is not limited to one direction so it is isotropic. Diffusion-weighted imaging (DWI) is based on the Brownian motion principle that allowing quantification and restriction of the free diffusing water molecules in tissues creates a phase that can be utilized in the imaging without the aid of contrasting agents (Chenevert et al., 1990; Doran et al. 1990). Diffusion-weighted MRI (DW-MRI) has been assessed for the generation of quantitative and early imaging biomarkers of therapeutic response and clinical outcome. As a highly sensitive tool for the detection of changes in the cellular level in most solid tumors, DW-MRI is based on the measurement of an increase in the apparent diffusion coefficient (ADC) of water molecules within the cancer lesion. In cancer treatment management, the incorporation of quantitative DW-MRI can help clinicians to individualize therapy, which minimizes unnecessary systemic toxicity associated with ineffective therapies, saves valuable time, reduces patient care costs, and finally improves the clinical outcome. DWI is used to detect cerebral ischemia, multiple sclerosis, brain tumors, and more (Baird & Warach, 1998; Galbán et al., 2017; Kono et al., 2001; Larsson et al., 1992; Moseley et al., 1990; Stadnik et al., 2001).

4.8.4 Arterial spin labeling (ASL)

ASL is a method of assessing tissue perfusion without the use of an exogenous contrast tracer. It is based on the principle that protons in the arterial blood water act as a natural endogenous tracer to aid in the assessment of tissue perfusion. In this process, first control images of tissue of interest are acquired, and then the protons in the blood flow are labeled before entering into the tissue by inverting the magnetization. After a time gap, images are acquired, and control images are subtracted from the labeled images, which generate the perfusion-weighted image. This image further undergoes processing to generate a cerebral blood flow map. ASL is used to assess and quantitate cerebral arterial territories in cerebrovascular diseases. Similarly, it is applicable in acute stroke and chronic cerebrovascular disease for the visualization of the collateral blood supply in the penumbra and the extent and severity of compromised cerebral perfusion, respectively, which help guide therapeutic or preventive intervention. ASL is also helpful in the detection and follow-up of arteriovenous malformations, brain tumors, epilepsy, neurodegenerative disease, renal imaging, pancreas imaging, and placenta imaging (Grade et al., 2015; Roberts et al., 1995; Taso et al., 2018).

4.8.5 Neuromelanin-sensitive MRI

Neuromelanin is a catecholamine-based polymer pigment rich in the specific cholinergic neurons of certain areas of the brain. Substantia nigra in the midbrain and locus coeruleus in pons and nucleus tractus solitarius to a certain extent are rich in neuromelanin-concentrated neurons. When neuromelanin chelates with iron, it behaves like a paramagnetic substance, which provides good contrast with magnetization transfer. Due to magnetization transfer, the neuromelanin-rich area appears with much signal intensity in T_1-weighted images and distinguishes it from the background. Neuromelanin-sensitive MRI can be used as a noninvasive proxy measure of dopamine function in the human brain (Cassidy et al., 2019; Sasaki et al., 2008).

4.8.6 Quantitative susceptibility mapping

Each tissue has magnetic susceptibility similar to the T_1 and T_2 relaxation constants, and the research was beginning to use this susceptibility information in imaging studies by using the phase. Susceptibility weighted imaging (SWI) is aiding the quantitative susceptibility mapping in using the phase that enhances the T_2 contrast. In clinical practice, phase contrast in tissues is provided by iron, lipid, calcium, and myelin content. Iron provides the phase contrast in the diagnosis of parkinsonism as well in the patients with suspected pathology resulting from microhemorrhages, and deoxyhemoglobin provides phase contrast in assessing the blood flow changes. To characterize brain tissue, SWI is based on a high-resolution, full velocity-compensated, 3D gradient-echo sequence using magnitude and phase images in combination with each other or either separately. The SWI is useful in quantification of leptomeningeal collateralization in patients with acute cerebral infarction and in the setting of trauma and acute neurologic presentations suggestive of a stroke. SWI can also characterize neurodegenerative diseases, occult low-flow vascular malformations, intracranial calcifications, cerebral microbleeds, and brain tumors (Halefoglu & Yousem, 2018; Yang & Luo, 2018).

4.9 Current clinical applications of MRI in cardiovascular diseases

Cardiac MRI is a versatile, important, and accurate tool for multiparametric morphologic and functional evaluation in cardiovascular diseases. MRI applies in determining acute and chronic ischemic cardiac disease to the examination of the substrate of complex ventricular arrhythmias and the follow-up of patients with valvular and congenital heart disease. Cardiac MRI is also an invaluable technique for the accurate detection, diagnosis, characterization, and follow-up of several cardiac diseases in patients in terms of cardiac volume and ejection fraction quantification, cardiac shunt assessment, tissue characterization, valvular regurgitant fraction and pharmacologic stress myocardial perfusion, and 3D reconstruction of great vessels (Pedrotti et al., 2006).

4.9.1 Assessment of left and right ventricular volumes and mass, as well as systolic function

Cardiovascular MRI gives many more accurate values of left ventricular parameters than planar imaging modalities such as echocardiogram or ventriculography, especially in conditions such as ischemic or dilated cardiomyopathy where the geometry of the heart is altered. Cardiovascular MRI is the best imaging modality for patients undergoing therapeutic intervention. It gives information about the ventricular volumes if the patient is planned for pulmonary valve replacement in operated tetralogy patients (Kuehne et al., 2004).

4.9.2 Assessment of myocardial viability and myocardial perfusion

The decision to take myocardial revascularization is based on the viability of the myocardium (improvement of the contraction after revascularization). Myocardial fibrosis, stunning, and hibernation are best differentiated with MRI. Delayed contrast enhancement with gadolinium can differentiate the areas of acute and chronic infarction and edema (Medical Advisory Secretariat, 2010).

4.9.3 Evaluation of congenital heart disease with shunt calculation

It is a useful tool in the serial follow-up of patients with congenital heart diseases especially in operated patients. It can be used to identify congenital lesions such as atrial septal defect (ASD), ventricular septal defect (VSD), patent ductus arteriosus (PDA), and total anomalous pulmonary venous return (TAPVC). It can be used to calculate the shunt ratio with velocity mapping (Choe et al., 2001; Dorfman & Geva, 2006).

4.9.4 Evaluation and follow-up of valvular disease

Although echocardiogram is the investigation of choice to interrogate the valve, MRI is complementary to valvular disease especially in the follow-up (Pedrotti et al., 2006).

4.9.5 Evaluation of pericardial disease

Pericardial thickening and effusion are best visualized with MRI. Salient diagnostic criteria for a pericardial disease such as pericardial thickness, atrial dilatation, and diastolic septal bounce are better visible with MRI.

4.9.6 Evaluation of aortic disease

Three variants of acute aortic syndrome (e.g., aortic dissection, intramural hematoma, and penetrating aortic ulcer) are best diagnosed with MRI. Aortic aneurysm and its relation to its branch vessels are well defined with MRI. Contrast-enhanced MRI is a useful tool in diagnosing aortic dissection and its extent. Cardiac MRI is also useful in follow-up after aortic dissection or aneurysm surgery and in the diagnosis of coarctation.

4.9.7 Evaluation of cardiac masses

Transthoracic echocardiogram is useful in assessing the cardiac masses. MRI is a complementary imaging modality to echocardiogram; it can more reliably demonstrate the location and extent of the mass. MRI can give information about the nature of the tumor to some extent, although pathologic evaluation can establish a definitive diagnosis.

4.9.8 Cardiomyopathies

Cardiovascular MRI can diagnose hypertrophic cardiomyopathy and its functional squeal dynamic left ventricular outflow obstruction, regurgitation. Compared to the echocardiogram, MRI is more precise in diagnosing apical septal hypertrophy. When a contrast agent is given, the degree of enhancement correlates with the risk of sudden death, the presence of left ventricular dilation, and heart failure. It is useful in assessing the effect of percutaneous septal ablation and its functional squeal. Contrast-enhanced MRI can differentiate between the ischemic and nonischemic variety of dilated cardiomyopathy. Based on ventricular parameters it provides the prognostic information.

4.9.9 Arrhythmogenic right ventricular cardiomyopathy

It is characterized by the replacement of left and right ventricular myocardium with fibrofatty tissue in young individuals. Although the diagnosis is made on clinical grounds, being its prolonged silent phase, MRI aids in the early diagnosis of the arrhythmogenic right ventricular cardiomyopathy before the symptoms become more severe.

4.10 Role of MRI in diagnosis, staging, and disease evaluation in cancers

MRI aids an important role in diagnosis, staging, and disease evaluation in endometrial cancer (EC) and cervical cancer (CC). As per the International Federation of Gynecology and Obstetrics (FIGO) criteria, MRI explains myometrial invasion depth, which corresponds with tumor grade and lymph node metastases and thus associates with prognosis. MRI can accurately determine prognostic markers such as tumor size, pelvic sidewall, parametrial invasion, and lymph node invasion. However, some challenges still exist in the correct and accurate diagnosis of EC and CC. Hence, multiple approaches for the correct and accurate diagnosis of EC and CC have been required, which include functional MRI, DWI and Dynamic contrast-enhanced (DCE) sequences, biopsy, and histopathologic analysis (Otero-García et al., 2019). Recent advancements in the applications of PET/MRI hybrid imaging technology in clinical oncology to improve the diagnostic evaluation of various types of cancers in routine clinical use are evolving to get the high-resolution anatomic data and functional imaging data from MRI and PET, respectively (Sotoudeh et al., 2016).

4.11 Contraindications and limitations of MRI

MRI has no ionizing effect on tissues; no long-term adverse effects have been documented. Although some metallic implants such as modern cardiac valves, coronary stents, and sternal wires produce local artifacts, they do not pose any hazard to the patient. Patients with implanted defibrillators and pacemakers should not undergo MRI. Patients with neurostimulators, cochlear implants, vascular clips, and metal fragments retained in the eye should not even enter into the scan station. Pregnant ladies can undergo MRI if the benefits overweight the theoretic risk of embryopathy. However, intrauterine contraceptive devices (made of plastic or copper material) do not interfere with the magnetic field, do not produce heat and major artifacts (at the medium or high magnetic field), and such patients can be imaged safely with MRI. Nearly 5% of patients who undergo MR scan experience claustrophobia that necessities mild sedation. Movement of the patient while acquiring the scan produces the artifacts so patient cooperation is necessary. Uncooperative patients and children less than 10 years of age usually require conscious sedation to complete MR examination to avoid motion artifacts. Although gadolinium can be administered safely in adults and children of more than 6 months, nephrogenic systemic fibrosis (NSF) is a rare complication seen in patients with renal insufficiency (GFR <30 mL/min/1.73 m^2) manifested by skin symptoms, diffuse fibrosis of skeletal muscle, bones, lungs, pleura, pericardium, myocardium, kidney, testes, and dura. However, MRI has a major advantage over other imaging techniques in that it can be used to image a specimen/object in multiple planes such as axial, oblique, coronal, and sagittal planes.

4.12 Conclusion

In recent years, extensive progress has been made for the advancement of clinical MRI. In 2016, a multicontrast MRI system with MAGiC (MAGnetic resonance image Compilation) software (GE Healthcare) was approved by the US Food and Drug Administration (FDA) that enables acquisition of eight image contrasts (e.g., T_1, T_2, T_1-FLAIR, T_2-FLAIR, short τ inversion recovery [STIR], dual inversion recovery [DIR], phase-sensitive inversion recovery [PSIR], and proton density-weighted images of the brain) in a single acquisition within a fraction of time. The introduction of the Ultrashort Echo Time (UTE) sequence imaging system in 2015 (Toshiba) provided the choice to use the system dedicated to pulmonary MRI as it overcomes the complexities associated with lung MRI (e.g., requirement of the low-density hydrogen nuclei/atoms) as it is filled with air. Recent introduction of an advanced MRI system (Philips) powered with new ScanWise implant (implant-specific setup) software that offers automated user interface technology, the limitations associated with medical implants have been resolved to a larger extent. The 7-T MRI systems are generally limited to advance research use, but in 2017 the FDA approved Magnetom Terra (Siemens Healthineers), a 7-T MRI system for clinical use (neurologic and musculoskeletal imaging) in the United States. The major limitation of the MRI system and its technologic advancement is the high cost of the facility. There are many efforts being made to develop low-cost MRI systems so that in the future the MRI systems will be readily available in small clinical setup particularly in developing countries.

References

Avdievich, N.I., 2011. Transverse electromagnetic (TEM) coils for extremities. Journal of Magnetic Resonance https://doi.org/10.1002/9780470034590.emrstm1127.

Baird, A.E., Warach, S., 1998. Magnetic resonance imaging of acute stroke. Journal of Cerebral Blood Flow and Metabolism 18 (6), 583–609.

Bloch, F., Hansen, W.W., Packard, M.E., 1946. Nuclear induction. Physical Review 69, 127.

Carr, D.H., Brown., J., Bydder, G.M., et al., 1984. Intravenous chelated gadolinium as a contrast agent in NMR imaging of cerebral tumours. Lancet 1, 484–486.

Cassidy, C.M., Zucca, F.A., Girgis, R.R., Baker, S.C., Weinstein, J.J., Sharp, M.E., Bellei, C., Valmadre, A., Vanegas, N., Kegeles, L.S., Brucato, G., Kang, U.J., Sulzer, D., Zecca, L., Abi-Dargham, A., Horga, G., 2019. Neuromelanin-sensitive MRI as a noninvasive proxy measure of dopamine function in the human brain. Proceedings of the National Academy of Sciences of the United States of America 116 (11), 5108–5117.

Chenevert, T.L., Brunberg, J.A., Pipe, J.G., 1990. Anisotropic diffusion in human white matter: Demonstration with MR techniques in vivo. Radiology 177 (2), 401–405.

Choe, Y.H., Kang, I.S., Park, S.W., Lee, H.J., 2001. MR imaging of congenital heart diseases in adolescents and adults. Korean Journal of Radiology 2 (3), 121–131.

Corea, J.R., Flynn, A.M., Lechene, B., et al., 2016. Screen-printed flexible MRI receive coils. Nature Communications, 7.

Di Costanzo, A., Trojsi, F., Tosetti, M., et al., 2003. High-field proton MRS of human brain. European Journal of Radiology 48 (2), 146–153.

Doran, M., Hajnal, J.V., Van Bruggen, N., King, M.D., Young, I.R., Bydder, G.M., 1990. Normal and abnormal white matter tracts shown by MR imaging using directional diffusion-weighted sequences. Journal of Computed Assisting Tomography 14 (6), 865–873.

Dorfman, A.L., Geva, T., 2006. Magnetic resonance imaging evaluation of congenital heart disease: Conotruncal anomalies. Journal of Cardiovascular Magnetic Resonance: Official Journal of the Society for Cardiovascular Magnetic Resonance 8 (4), 645–659.

Galbán, C.J., Hoff, B.A., Chenevert, T.L., Ross, B.D., 2017. Diffusion MRI in early cancer therapeutic response assessment. NMR in Biomedicine 30 (3). doi:10.1002/nbm.3458.

Grade, M., Hernandez Tamames, J.A., Pizzini, F.B., Achten, E., Golay, X., Smits, M., 2015. A neuroradiologist's guide to arterial spin labeling MRI in clinical practice. Neuroradiology 57 (12), 1181–1202.

Grover, V.P., Tognarelli, J.M., Crossey, M.M., Cox, I.J., Taylor-Robinson, S.D., McPhail, M.J., 2015. Magnetic resonance imaging: Principles and techniques: Lessons for clinicians. Journal of Clinical and Experimental Hepatology 5 (3), 246–255.

Halefoglu, A.M., Yousem, D.M., 2018. Susceptibility weighted imaging: Clinical applications and future directions. World Journal of Radiology 10 (4), 30–45.

Hawkes, R.C., Holland, G.N., Moore, W.S., Worthington, B.S., 1980. Nuclear magnetic resonance (NMR) tomography of the brain: A preliminary clinical assessment with demonstration of pathology. Journal of Computed Assisting Tomography 4 (5), 577–586.

Junge, S., 2012. Cryogenic and superconducting coils for MRI. Journal of Magnetic Resonance, 1.

Keil, B., Blau, J.N., Biber, S., et al., 2013. A 64-channel 3T array coil for accelerated brain MRI. Magnetic Resonance Medicine 70, 248–258.

Kono, K., Inoue, Y., Nakayama, K., et al., 2001. The role of diffusion-weighted imaging in patients with brain tumors. American Journal of Neuroradiology 22 (6), 1081–1088.

Kuehne, T., Yilmaz, S., Steendijk, P., Moore, P., Groenink, M., Saaed, M., Weber, O., Higgins, C.B., Ewert, P., Fleck, E., Nagel, E., Schulze-Neick, I., Lange, P., 2004. Magnetic resonance imaging analysis of right ventricular pressure-volume loops: In vivo validation and clinical application in patients with pulmonary hypertension. Circulation 110 (14), 2010–2016.

Labbé Atenas, T., Ciampi Díaz, E., Cruz Quiroga, J.P., Uribe Arancibia, S., Cárcamo Rodríguez, C., 2018. Functional magnetic resonance imaging: Basic principles and application in the neurosciences. (Resonanciamagnéticafuncional: principiosbásicos y aplicacionesenneurociencias.). Radiologia 60 (5), 368–377.

Larsson, H.B., Thomsen, C., Frederiksen, J., Stubgaard, M., Henriksen, O., 1992. In vivo magnetic resonance diffusion measurement in the brain of patients with multiple sclerosis. Magnetic Resonance Imaging 10 (1), 7–12.

Lauterbur, P.C., 1973. Image formation by induced local interactions: Examples of employing nuclear magnetic resonance. Nature 242 (5394), 190–191.

Liu, C.H., 2015. Anatomical, functional, and molecular biomarker applications of magnetic resonance neuroimaging. Future Neurology 10 (1), 49–65.

Secretariat, Medical Advisory, 2010. Magnetic resonance imaging (MRI) for the assessment of myocardial viability: An evidence-based analysis. Ontario Health Technology Assessment Series 10 (15), 1–45.

Meijer, F., Goraj, B., Bloem, B.R., Esselink, R., 2017. Clinical application of brain MRI in the diagnostic work-up of parkinsonism. Journal of Parkinson's Disease 7 (2), 211–217.

Mitterschiffthaler, M.T., Ettinger, U., Mehta, M.A., Mataix-Cols, D., Williams, S.C., 2006. Applications of functional magnetic resonance imaging in psychiatry. Journal of Magnetic Resonance Imaging 23 (6), 851–861.

Moseley, M.E., Kucharczyk, J., Mintorovitch, J., et al., 1990. Diffusion-weighted MR imaging of acute stroke: Correlation with T2 weighted and magnetic susceptibility-enhanced MR imaging in cats. American Journal of Neuroradiology 11 (3), 423–429.

Nordmeyer-Massner, J.A., De Zanche, N., Pruessmann, K.P., 2012. Stretchable coil arrays: Application to knee imaging under varying flexion angles. Magnetic Resonance Medicine 67, 872–879.

Otero-García, M.M., Mesa-Álvarez, A., Nikolic, O., Blanco-Lobato, P., Basta-Nikolic, M., de Llano-Ortega, R.M., Paredes-Velázquez, L., Nikolic, N., Szewczyk-Bieda, M., 2019. Role of MRI in staging and follow-up of endometrial and cervical cancer: Pitfalls and mimickers. Insights into Imaging 10 (1), 19.

Pedrotti, P., Pedretti, S., Imazio, M., Quattrocchi, G., Sormani, P., Milazzo, A., Quarta, G., 2006. Risonanzamagneticacardiaca: istruzioni per l'uso. Cardiopatiaischemica, miocardite, malattie del pericardio, aritmie, valvulopatie, cardiopatiecongenite e masse cardiache [Clinical applications of cardiac magnetic resonance imaging: coronary heart disease, myocarditis, pericardial diseases, arrhythmias, valvular heart disease, congenital heart disease and cardiac masses. Giornaleitaliano di Cardiologia 20 (1), 8–19.

Purcell, E.M., Torrey, H.C., 1946. Pound RV. Resonance absorption by nuclear magnetic moments in a solid. Physical Review 69, 37–38.

Requardt, H., Offermann, J., Kess, H., et al., 1987. Surface coil with variable geometry: A new tool for MR imaging of the spine. Radiology 165, 572–573.

Roberts, D.A., Detre, J.A., Bolinger, L., Insko, E.K., Lenkinski, R.E., Pentecost, M.J., Leigh Jr, J.S., 1995. Renal perfusion in humans: MR imaging with spin tagging of arterial water. Radiology 196 (1), 281–286.

Sasaki, M., Shibata, E., Kudo, K., et al., 2008. Neuromelanin-sensitive MRI. Clinical Neuroradiology 18, 147–153.

Slavkovsky, P., Uhliar, R., 2004. The Nobel prize in physiology or medicine in 2003 to Paul C. Lauterbur, Peter Mansfield for magnetic resonance imaging. Bratislavskelekarskelisty 105 (7-8), 245–249.

Smith, F.W., Hutchison, J.M., Mallard, J.R., et al., 1981. Oesophageal carcinoma demonstrated by whole-body nuclear magnetic resonance imaging. British Medical Journal (Clin Res Ed) 282 (6263), 510–512.

Soher, B.J., Dale, B.M., Merkle, E.M., 2007. A review of MR physics: 3 T versus 1.5 T. Magnetic Resonance Imaging Clinics of North America 15 (3), 277–290.

Sotoudeh, H., Sharma, A., Fowler, K.J., McConathy, J., Dehdashti, F., 2016. Clinical application of PET/MRI in oncology. Journal of Magnetic Resonance Imaging 44 (2), 265–276.

Stadnik, T.W., Chaskis, C., Michotte, A., et al., 2001. Diffusion-weighted MR imaging of intracerebral masses: Comparison with conventional MR imaging and histologic findings. American Journal of Neuroradiology 22 (5), 969–976.

Taso, M., Guidon, A., Zhao, L., Mortele, K., Alsop, D., 2018. Pancreatic perfusion and arterial-transit-time quantification using pseudocontinuous arterial spin labeling at 3T. *Magnetic Resonance in Medicine*.

Westbrook, C., Roth, C.K., Talbot, J., 2011. MRI in practice, 4th ed. John Wiley & Sons, Inc.

Wiggins, G.C., Brown, R., Lakshmanan, K., 2016. High-performance RF coils for 23Na MRI: Brain and musculoskeletal applications. NMR Biomedicine 29, 96–106.

Wright, A.C., Lemdiasov, R., Connick, T.J., et al., 2011. Helmholtz-pair transmit coil with integrated receive array for high-resolution MRI of trabecular bone in the distal tibia at 7 T. Journal of Magnetic Resonance Medicine 210, 113–122.

Yang, L., Luo, S., 2018. Clinical application of susceptibility-weighted imaging in the evaluation of leptomeningeal collateralization. Medicine 97 (51), e13345.

Young, I.R., Clarke, G.J., Bailes, D.R., et al., 1981. Enhancement of relaxation rate with paramagnetic contrast agents in NMR imaging. Journal of Computed Tomography 5, 543–547.

Current update about instrumentation and utilization of PET-CT scan in oncology and human diseases

Rohit Gundamaraju[a], Chandrabhan Rao[b], Naresh Poondla[c]

[a]*ER stress and Mucosal Immunology Team, School of Health Sciences, University of Tasmania, Launceston, Tasmania, Australia*
[b]*Abhinn Innovation Private Limited, Jaipur, Rajasthan, India*
[c]*Richmond University Medical Center, 355 Bare Avenue, Staten Island, NY, United States of America*

5.1 Introduction

Diagnosis stands as the prime rescuer in numerous diseases, especially in cancers. Imaging techniques stand as critical decision-making tools for making evaluations such as grading of disease and staging in cancers. Recent advances in imaging techniques such as colonoscopy, computed tomography (CT), advanced magnetic resonance, and novel organ imaging are storming the medical industry aiding in a better understanding of the disease. Image-based analysis of tumor heterogeneity and multiparametric imaging, the development of radiomics and radiogenomics, and more, are utmost necessary in diseases such as colorectal cancer (CRC), breast cancer, and brain tumors because of their aggressive nature and prevalence.

As an integral part of nuclear medicine imaging, positron emission tomography (PET) makes quantitative in vivo measurements of three-dimensional (3D) positron-emitting tracer distributions. The most popular and widely used PET tracer in oncology applications is ^{18}F-fluorodeoxyglucose (FDG). Basically, FDG offers a glucose intake indicator and helps in identifying several different forms of tumors (Sattler et al., 2010). Different physiologic or pharmacokinetic parameters, such as blood flow, glucose, and oxygen consumption, neuroreceptor density and affinity, drug delivery and uptake, and gene expression, may be derived by using various tracers. Furthermore, PET can be used as a clinical application to determine therapeutic responses or to assess the efficacy of new drugs. High-resolution standards (now up to ~2.5 mm full width at half-maximum for clinical PET scanners) are integrated into PET imaging. The visual analysis of PET photographs offers enough detail for some of these applications. For tumor staging and patient treatment, visual analysis of

Biomedical Imaging Instrumentation. **DOI: https://doi.org/10.1016/B978-0-323-85650-8.00007-3**

whole-body FDG images is normally used in oncology. The advancement of exciting positron-based radiopharmaceutic technology has contributed to the development of PET instrumentation over the past 40 years (Fletcher et al., 2008).

5.2 PET imaging and instrumentation

Most PET systems are PET-CT multimodality systems that are integrated. Both PET-CT systems have a sequential but integrated device design in which the CT scanner is located either within a wide cover or in two separate covers in front of the PET component, enabling the two systems to be moved apart. A whole-body PET and a normal, clinical multislice CT within a single portal are integrated with all PET-CT systems for clinical use.

5.2.1 Principles of PET

PET is a technique for molecular imaging that measures in vivo the distribution of a radioactive tracer. The decay mechanism of positron emission is characterized by the conversion of a protein into a positron-emitting neutron, the electron's positively charged antiparticle, and a neutrino, a particle that is chargeless and almost massless and does not interact with matter. A pair of two photons (γ-rays) is produced when an emitted positron unites with a nearby electron in an annihilation event, each having energy of 511 keV, which is emitted in nearly opposite directions. To detect these pairs of photons when they strike opposing detectors at about the same time, PET scanners are fitted with coincidence (γ-ray) detectors. Millions of coincidence detections are obtained during a PET scan, providing data on the tissue distribution of the radiotracer. Unfortunately, not all coincidences lead to the signal, and background noise is introduced to the signal due to photons that are spread until two uncorrelated photons are detected or coincidentally detected (Spanoudaki & Levin, 2010). Until leaving the patient, a significant fraction of the released photons (up to 50%) is distributed, resulting in a dislocation of the true detection of coincidence. Moreover, the PET camera will find spontaneous coincidence detection when two photons from two separate positron emissions are randomly detected simultaneously (while the others are undetected). Finally, when three or more photons are detected at the same time, multiple detections may occur.

Quantifying PET studies involves accounting for the contributions of scattered and random coincidences. The attenuation of PET does not depend on the position of the positron emission along the response line. Consequently, the results of attenuation may be reliably reversed by receiving transmission and or CT scans (Wolk et al., 2012).

5.2.2 Acquisition and image reconstruction: 2D versus 3D

Although PET is a 3D imaging technique, many PET and PET-CT scanners have been equipped with septa (i.e., lead or tungsten annular shields positioned in the field of view [FOV]) in the past. Compared to 3D (no septa) acquisitions, the primary aim of using these septa (2D mode) was to reduce the contribution of random coincidences, scattered photons, and photons coming from the operation. Most of the current PET and PET-CT scanners are no longer fitted with septa, and the only alternative is 3D acquisition (without septa). 3D acquisitions have a higher detection rate, leading to increased sensitivity but also in increased contributions from random and scatter. These new crystal materials result in better performance of the count rate as they demonstrate a faster rise and decay time of scintillation. As a consequence, it is possible to apply a shorter coincidence time window, resulting in a decrease in random coincidences (and scatter) (Chapman et al., 2012).

5.2.3 **Random correction**

While the coincidence window easily excludes single unpaired photons, in PET imaging additional processes occur that degrade the image contrast. The first is called random, which happens by coincidence when single photons from two separate annihilations are observed. The random rate often exceeds the true coincidence rate in whole-body PET imaging; however, the random rate rarely exceeds 25% of the true rate in brain imaging. The second method is scatter, in which either or both photons are deflected from their usual course from extinction event but maintain enough energy to be accepted by the detection device. Both of these processes allow the event of a coincidence to be allocated to the wrong line of response (LOR), thereby misrepresenting the true distribution of operation within the field of view (Garibotto et al., 2013). Thus, during data processing, different algorithms and techniques are used to compensate for random and scatter.

5.2.4 **Attenuation correction**

To produce a final image, several extra corrections are needed during data processing. It is relatively straightforward for brain imaging to compensate for dead time, a count rate–dependent phenomenon by which the detector is paralyzed and stops recording counts, and for varying detector sensitivity. For accurate quantitative results, the attenuation correction is one of the most significant corrections to the acquired PET image (Rottenburger et al. 2011). This correction is required because photons emitted from the center of the brain are attenuated to a greater degree than photons released from the periphery, and the strength in the center of the brain is reduced by around 50% without attenuation correction. PET data attenuation correction is easier than SPECT data correction since, unlike SPECT, at the endpoint along a LOR, the likelihood of detection is uniform.

Finally, the FOV of the CT is smaller than that of the PET scanner in some PET-CT scanners. If the patient is only partly visible on the CT scan (e.g., the patient's arms are often truncated), it is not possible to measure the attenuation accurately for certain LORs going through the arms, and attenuation correction may be inaccurate.

5.2.5 Scatter correction

When either or both photons are dispersed and thus deflected from their original path, scattered coincidences occur. Compton scattering is the primary source of scattering events for 511-keV photons, where the photon interacts with (or "hits") an electron. The photon has lower energy after scattering (depending on the angle of deflection from its original direction), and a different direction is obtained. Using a single scatter simulation method, the most widely used scatter correction method in PET is based on estimating the scatter distribution/contribution. Since the technique is based on the physical concepts of photon scattering and takes account of operation and attenuation distribution, it appears to be an effective method of scatter correction and is now routinely available (Huang et al., 2009).

5.2.6 Image reconstruction methods

PET measures coincidences that are commonly found in sinograms. A sinogram includes the estimates of the activity distribution in the patient over all angles. The method of measuring the patient's 3D activity distribution from the measured sinograms, including random, scatter, attenuation, normalization, and dead time correction, is called image reconstruction. Filtered back-projection (FBP) is the most widely used analytic image reconstruction technique. This is a linear and quantitatively stable methodology. The system is vulnerable to noise, however, and reconstructed images contain extreme artifact of streaks. Iterative reconstruction algorithms, such as ordered subset expectation maximization (OSEM), have been developed for these purposes. Other iterative methods such as less square and row-action maximum likelihood algorithm (RAMLA) have also been developed. However, insistent reconstruction can display biases (i.e., quantitative inaccuracies) in the case of complex PET studies consisting of several frames with short scan durations and thus weak statistics. Consequently, for complex PET research, FBP is also still the preferred reconstruction technique (Varrone et al., 2009). Therefore modern PET-CT systems are equipped with dedicated computer clusters to recreate the image within a reasonable period of time, and the quality of fully 3D reconstruction has become the standard.

5.2.7 The need for PET screening in diverse cancers

Cancer is a global health menace of which CRC represents the second leading cause of death by cancer (Ferlay et al., 2015). Besides diet and aging, risk factors such as smoking and obesity escalate the risk of CRC. A deep understanding of the pathophysiology increased the ray of hope in treatment options and regimens. Current treatments for CRC include excisions, pre- and postoperative radiotherapy, surgeries, chemotherapy, palliative care, and targeted immune therapies. Speaking of pathology, it is estimated that 2% to 5% of adenomas lead to CRC. Despite huge development in treatment options, the majority of patients die of metastatic disease. The fact that supports the notion of its lengthy transition from early to symptomatic disease in CRC provides an excellent window for screening (Kuipers et al., 2013).

In recent times, a high volume of diagnostic challenges has been answered by PET especially in neurocancers. In brain cancer detection, PET images are useful in greater detection of tumors such as glioma where there was a mean sensitivity of 82%. Considering the inability of conventional screening methods like MRI to identify some of the categories of glioma-like nonenhancive, radiolabeled amino acids, paired PET-CT is employed since it can cross the blood-brain barrier. Even with regard to the volumetric comparison, studies have displayed significant success (Galldiks et al., 2019a).

Into the bargain, lung cancer has been observed as one of the most common causes of death in both sexes. Since the vast majority of radiographically indeterminate nodules are benign, it is clear that there is a serious need for an accurate diagnosis. PET-CT has greatly aided in the evaluation of lung cancers by allowing better delineation of areas with an enhanced tracer uptake. Moreover, it has been an excellent tool for solitary pulmonary nodules and detailed lung cancer screening. In various instances, a 100% detection of lesions was obtained by virtual 3D images versus axial images and demonstrated subjective improvements. These improvements have been possible by fusing 3D modeling with PET-CT, which massively improves detection capabilities (Bunyaviroch & Coleman, 2006).

On the other hand, breast cancer attributes to over 40,000 deaths in the United States alone making it one of the most prevalent diseases in women. Though mammography, ultrasound, and magnetic resonance imaging (MRI) can help spot local disease, imaging screenings such as PET-CT and FDG PET-CT play a key role in systemic staging in specific patients. The National Comprehensive Cancer Network (NCCN) suggested that PET-CT can be utilized for stage III ideally but not for I and II because of high levels of false positives. However, PET-CT is highly handy in detecting extraaxial nodal disease in almost 25% of the cases and in finding out distant metastasis in 10% to 20% of the cases, which is regarded utmost significant (Bellon et al., 2004; Fuster et al., 2008; Mahner et al., 2008).

5.3 Clinical significances of PET in cancer

Considerable assistance is necessary for clinicians in cancer management. PET-CT colonography in the CRC detection has been a hit with a sensitivity of nearly 95%, whereas a sensitivity of 97% to 100% was achieved in lung carcinoma (Hochhegger et al., 2015). The sensitivity in breast cancer was recorded as 68% for small (<2 cm) tumors and 92% for larger (2–5 cm) tumors (Yang et al., 2007). Fig. 5.1 is a classic example of the detection of malignancy. The malignant cells have taken the dye and are eluted. This gives a clear idea about cancer. Accurate

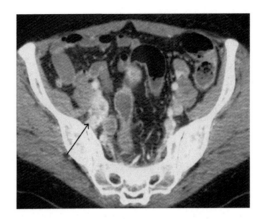

FIG. 5.1

Uptake of FEG-PET and a lesion observed on CT, which is supposed to be a malignant deposit. (From O'Connor, O. J., McDermott, S., Slattery, J., Sahani, D., & Blake, M. A. [2011]. The use of PET-CT in the assessment of patients with colorectal carcinoma. *International Journal of Surgical Oncology*, 2011846512.).

preoperative staging is utmost necessary in the prognosis and in providing the required therapy in cancers like CRC. The first type of staging is T staging. FDG-PET was found inappropriate for local staging due to its limited spatial resolution. PET was regarded as accurate in local invasions. The T staging of CRC is entirely dependent on CT.

The suitability and sensitivity of the technique or the equipment vary vastly for the different T staging (e.g., the imaging of T3 and T4 tumors is better performed with either MRI or CT). Distant metastasis is regarded as a tough spotting in imaging. Multislice CT is the best choice for mesorectum and multidistant metastasis screening with a sensitivity of 74% and specificity of 95%. N staging remains important, too. PET-CT is primarily utilized in determining treatment approach and can also aid in altering staging mainly due to revised nodal staging (O'Connor et al., 2011).

5.3.1 Clinical endorsement by PET-CT

PET-CT has been put to use for the initial staging of CRC widely. In various studies, the sensitivity and specificity of PET-CT have been well documented (Agarwal et al., 2014). In 2010, a study documented the tumor invasion beyond muscularis propria with a sensitivity and specificity of 86% and 78%, respectively, whereas the sensitivity and specificity values of nodal detection were 70% and 78%, respectively (Dighe et al., 2010). PET-CT was also useful in determining bone metastasis and metastatic lesions in patients thereby leading to alterations in patient management (i.e., from chemotherapy to surgery). The accuracy of diagnostics for N staging with PET-CT was 66%. The diagnostic accuracy for T staging was recorded as

82% (Kunawudhi et al., 2016). This study suggested promising results in the initial staging of CRC. Local reoccurrence was also regarded as an important aspect to be considered in CRC management. FDG PET-CT showed a demonstrated sensitivity in assessing local reoccurrence and distant metastasis in a patient with pathologically detected CRC with a sensitivity and specificity of 95.45% and 96.7%, respectively, in a study consisting of 60 patients. Surprisingly, an accuracy of 100% was achieved in detecting hepatic metastasis and nodal metastasis (Hetta et al., 2020). In another meta-analysis, FDG PET was considered as the most accurate imaging technique for the discernment of liver metastasis in CRC (Bipat et al., 2005). Considering per-patient results, PET-CT achieved a startling sensitivity of 94.1% (Niekel et al., 2010). One of the difficult tasks in CRC treatment is the tumor response succeeding the chemotherapy. In a study by Georgakopoulos et al. (2013), FDG PET-CT detected extrahepatic disease, which was missed by conventional imaging. A meta-analysis of 3080 patients with CRC revealed the sensitivity of hepatic metastasis was 90% for FDG PET (Kinkel et al., 2002). The success of the PET-CT was assessed in a study by Selzner et al. (2004), where the failure rate of PET-CT was only in 11% of the cases. Kochhar and colleagues (2010) have evaluated the concept: upstaging and downstaging with the help of PET-CT where they have discovered a significant effect on management in CRC metastasis. Synchronous cancers, on the other hand, are defined as multiple distinct primary tumors separated by mucosa. The sensitivity of PET-CT for detecting synchronous invasive cancers in 61 patients was 66.6% thereby making it an effective tool in detecting synchronous colonic cancers in patients with obstructive colon cancer (Maeda et al., 2019). Well marginated, discrete, and surrounded by lung parenchyma is the solitary pulmonary nodule, which makes the job of conventional imaging tougher. F-FDG PET has enormously increased the predictive analysis of the disease. F-FDG uptake can easily differentiate benign nodules making it highly sensitive (97%) and specific (82%) (Nomori et al., 2004). Fig. 5.2 illustrates the positive employment of PET in cancer staging.

Notably, the nodules greater than 10 mm were detected with greater sensitivity by PET (Bastarrika et al., 2005). The anticipated value of F-FDG PET was determined by several studies where the detection improved mean survival time in patients (Ahuja et al., 1998; Downey et al., 2004).

PET-CT plays a prominent role in breast cancer patients where mammography is not desirable or biopsy is not recommended. Axillary lymph node metastasis is regarded as an important precursor in breast cancer. Axillary lymph node numbers are significantly correlated with active recurrence. Axillary PET imaging displays a sensitivity of 90% and specificity of 80% to 90%, which helps in accurately local-izing and differentiating metastatic and lymph nodes (Yang et al., 2007). PET-CT was advantageous over CT for internal mammary gland and mediastinal lymph node evaluation (Tatsumi et al., 2006).

PET images in brain tumor detection are highly useful, especially in glioma. A meta-analysis of nearly 400 patients indicated that MET PET showed a high pool sensitivity and specificity of 90% and 85%, whereas FDG PET yielded a moderate sensitivity of 70% (Zhao et al., 2014). In glioma detection, PET tracers like FET and

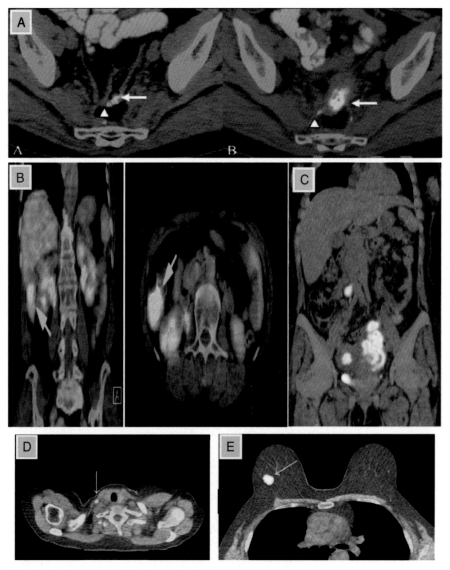

FIG. 5.2 Utilization of PET in pinpointing tumors.

(A) Coronal CT of an intraabdominal extrahepatic recurrence. (B) A lung tumor. (C) FDG PET-CT imaging revealing of an advanced colorectal neoplasm. (D) An axillary lymph node. (E) PET-CT of the liver. (A, From Taha Ali, T. F. [2012]. Usefulness of PET–CT in the assessment of suspected recurrent colorectal carcinoma. *The Egyptian Journal of Radiology and Nuclear Medicine, 43* [2], 129–137; C, Huang, S. W., Hsu, C. M., Jeng, W. J., Yen, T. C., Su, M. Y., & Chiu, C. T. [2013]. A comparison of positron emission tomography and colonoscopy for the detection of advanced colorectal neoplasms in subjects undergoing a health check-up. *PLoS One, 8* [7], e69111; E, Koolen, B. B., Vogel, W. V., Vrancken Peeters, M. J., Loo, C. E., Rutgers, E. J., & Valdés Olmos, R. A. [2012]. Molecular imaging in breast cancer: From whole-body PET/CT to dedicated breast PET. *Journal of Oncology*, 2012438647.).

FDOPA express high accuracy in differentiating progression and changes in treatment (Galldiks et al., 2019b).

There are numerous fundamental limitations of CT such as sensitivity, precise recognition of lymph nodes involved in the disease, and spotting of the extranodal tissue involved in the disease. PET-CT features greater sensitivity at such sites. PET-CT is preferred essentially in low-grade lymphoma. PET-CT assists greatly in baseline assessment of lymphoma in patients with stubborn or aggressive disease undergoing new line treatments (Cronin et al., 2010).

5.3.2 Detailed staging with the help of PET

In cancers like CRC, it is considered metastatic when the carcinoma is spread outside regional pericolic or mesenteric lymph nodes. Common sites of metastasis include lymph nodes, liver, bone, and tissues. PET-CT with contrast-enhanced portal venous phase was considered ideal for metastasis. Taking hepatic metastasis into account, PET-CT plays a prominent role in spotting additional extrahepatic sites leading to a successful treatment outcome. Combinational diagnosis with PET-CT and CT in a preoperative setting with liver metastasis leads to the identification of extrahepatic sites in a significant number of patients (Huebner et al., 2000; Strasberg et al., 2001). The high sensitivity of FDG PET-CT is shown in hepatic metastasis. A whopping 94% sensitivity was attained among other imaging techniques for liver metastasis (Niekel et al., 2010). Hence PET-CT can influence patient management by directing biopsies and directing surgical resections of liver metastasis. PET-CT can help find sites with carcinoembryonic antigen levels thereby aiding in the detection of recurrent or metastatic disease in patients with a history of CRC making it a better option over CT. The importance of PET-CT in hepatic metastasis was also proved by excluding nonsuitable or ineffective surgery in nonoperating disease. This made it cost efficient and a lifesaver by prolonging survival.

CRC spread to lymph nodes is named M1 disease or metastatic M1 (O'Connor et al., 2011). PET-CT is not highly recommended in detecting nodal metastasis because of factors such as FDG uptake, false negatives, and size or enlargement of nodes and detection. The lung is the second most common site for CRC metastasis per clinical statistics (O'Connor et al., 2011). Fortunately, the sensitivity and specificity of PET-CT for the detection of malignant pulmonary nodules are 96% and 86%. Hence PET-CT is the first choice to imagine technique resorted. Though conventional CTs can precisely identify nodules as minute as 2 to 3 mm it is still a puzzle between benign and malignancy because of specificity. Secondly, the distinction of metastasis becomes tougher in pulmonary and thoracic areas. FDG PET helps determine such abnormalities. With an increased standardized uptake value, sensitivity and specificity, FDG PET can more actively detect a smallest nodule than conventional CTs. PET-CT offers more effective gating for better spatial resolution. PET sensitivity also favors sites such as adrenal glands.

In lung cancer, the metastasis to lymph nodes is represented as N1. N2 disease denotes the spread to ipsilateral mediastinal lymph nodes. M is the representative alphabet for distant metastasis in lung cancer. Organs such as the brain, bone, liver, and adrenal glands are common sites for distant metastasis. [18]FDG-PET is regarded as sensitive for the detection of metastasis of adrenal glands. [18]FDG-PET aided in preventing surgery in at least one among five patients with non–small cell lung cancer (NSCLC) and making it an ideal scan procedure in mediastinal lymph node metastasis with a surprising sensitivity of 91% (Bunyaviroch & Coleman, 2006). The accuracy of PET scans in determining stages remains varied, especially the false negatives. However, PET was labeled significant over CT with regard to the mediastinal staging of lung cancer (Antoch et al., 2003; Lardinois et al., 2003).

FDG PET is extremely advantageous in breast cancer staging because of its ability in detecting unsuspected distant metastasis of the deep mammary lymph nodes, which are missed in the conventional screening. Early staging of breast cancer, sentinel lymph nodes screening is the right choice. A combination benefit was achieved with FDG PET and MRI for distinguishing axillary recurrence in patients. Moving to recurrences, FDG PET demonstrated superior abilities compared to CT or MR in a study of 75 patients (Bender et al., 1997), and high sensitivity and specificity were achieved in a study consisting of 57 patients (Moon et al., 1998). Further, FDG PET was noted to be a better choice over bone scintigraphy in detecting skeletal metastasis in breast cancer in 23 patients (Cook & Fogelman, 2000).

Similarly, PET-CT with contrast-enhanced MRI is immensely employable in N staging. High-grade or aggressive gliomas contained an obvious high FDG uptake. In astrocytic gliomas, (tumor to brain utilization ratio) patients were determined for their mean survival time with a specific ratio. Patients with ratios greater than 1:4:1 had a mean survival of 5 months, and ratios less than 1:4:1 had a mean survival of nearly 20 months. The majority of the tumors graded as I and II stage had a mean survival time of 2.5 years, whereas II and IV had a mean survival time of 11 months with a 94% grade. Gliomas that appear to be low grade are relatively tougher to detect by conventional diagnosis. FDG helps in grading the majority of surgically sorted malignant samples in glioma.

5.3.3 **PET-CT as support therapy**

CT is predominantly useful in determining planning target volumes before radiotherapy thereby recognizing the importance of FDG PET-CT in radiation planning. Before designing a radiation treatment, it would be important to reduce geographic misses or improve precision and reduction of nontumor spots being hit by radiation. Table 5.1 enumerates some of the endorsements to other therapies. Moreover, PET-CT aids in effectively covering nodal stations in gastrointestinal (GI) or CRC cancers. The following are some of the clinical studies showing the importance of PET-CT in a clinical setting of therapy planning or prior to radiation and therapy determination. In lung cancers like SCLC, [18]FDG-PET was compared with conventional radio imaging, and [18]FDG-PET changed the patient management for good. A median of 30% escalation was achieved. This allowed determining chemoradiation and chemotherapy in patients (Kamel et al., 2003).

Table 5.1 Various types of PET scans employed in various studies.

Imaging	Advantage attained	Study/references
PET-CT	Reduced the interobserver variability in the planning target volumes.	(Ciernik et al., 2003)
FDG PET-CT	Had a comparatively more similar index, which reflects interobserver variability among several radiation oncologists compared to CT alone.	(Patel et al., 2007)
PET-CT derived gross tumor volume (GTV)	The PET-CT derived GTV aided in adequate coverage of N1 lymph node stations in the perirectal soft tissues.	(Ciernik et al., 2005)
PET-CT	There was a pre- and postperformance of PET-CT for neoadjuvant therapy response (before chemotherapy and after a 7-wk period, which showed 61% positive PET-CT results with a specificity and sensitivity of 92.8% and 88.8%, respectively.	(Murcia Duréndez et al., 2013)
FDG PET	Accuracy of FDG PET in locally advanced CRC was recorded as more accurate than ultrasound with regard to chemotherapeutics response with a startling sensitivity of 100%.	(Amthauer et al., 2004)
PET-CT	Essential tool in early therapy assessment resulting in treatment management	(Avallone et al., 2012)
PET-CT	Responders have displayed an increased 5-year relapse-free survival, which in some responders also prompted for the usage of alternative therapies in case of early responders.	(Avallone et al., 2012)
PET-CT	Successful in assessing early response and relapse monitoring after radiofrequency ablation (RFA) therpy	(Israel & Kuten, 2007)

Likewise, in brain metastasis, patients undergoing targeted therapy showed progress measured via 3′-Deoxy-3′-[^{18}F]-fluorothymidine (FLT) PET. Similarly, amino acid PET also added valuable information alongside MRI scans (Galldiks et al., 2019b). Anatomic imaging has played a vital role in determining the state of the patient during chemotherapies in breast cancer. Smith et al. (2000) enumerated the FDG uptake levels before and after chemotherapy response by histologic examinations. PET prediction in a single-dose chemotherapeutic response was well documented by Rosé and colleagues (2002) with a sensitivity of 90%. A startling sensitivity of 100% was achieved in another study where all the patients were correctly identified who responded (Smith et al., 2000).

In diseases like lymphomas, PET-CT dictates the levels of residual lymphoma and hence is preferred as a posttreatment choice of the scan. PET-CT role was found to be vital in several studies where the patient's requirement for further treatment was alarmed. A pretreatment PET-CT is occasionally preferred depending on the refractory disease. The FDG uptake succeeding several cycles of chemotherapy has been directly proportional with poor clinical outcome, and chemotherapeutic regimens could be partially planned with PET-CT findings (Cronin et al., 2010).

5.4 Role of PET in the diagnosis of human diseases other than cancer

Currently, most (>95%) of the PET scans are used for oncology purposes. Over a decade, [^{18}F]FDG-PET was used as the gold standard in clinical diagnosis and research of human diseases, including cardiac, neurologic, and psychiatric diseases, among others. Applications of PET in cardiac diseases are based on some common cardiac PET tracers and their indicators such as ^{82}Rb-rubidium chloride, ^{13}N-ammonia, ^{15}O-water, ^{18}F-flurpiridaz, and ^{62}Cu-pyruvaldehyde bis(N-methylthiosemicarbazone) (PTSM) are utilized for myocardial perfusion and blood flow quantification. Moreover, cardiac PET tracers such as [^{18}F]FDG and ^{11}C-glucose are used for myocardial viability, inflammation, and glucose metabolism. These tracers/molecules can be utilized, quantified in miniscule amount (pico to femtomolar range) without affecting or damaging patient's vital activities, and imaged in vivo to obtain a specific signal with very high sensitivity during the PET scan for cardiac diseases (Bengel et al., 2009; Iqbal et al., 2014; Massoud & Gambhir, 2003; Welling et al., 2011). Thus cardiac PET can also be applicable when probing mechanisms with limited binding capacity observed. In cardiac PET, myocardial perfusion imaging (MPI) tests are the most widely used application (Klein et al., 2010). In the diagnosis of Alzheimer disease (AD), some clinical conditions such as obesity, atherosclerosis, and diabetes affect moderate patterns in parietotemporal hypometabolism in normal cognitive subjects that cause less than 95% specificity and limited sensitivity (84–96%) of [^{18}F]FDG-PET. Moreover, the availability of reliable markers and advancement of technologies (like MRI) in data acquisition and analysis has changed the diagnostic approach in neurologic disorders. The recent development in [^{18}F] FDG-PET in combination with amyloid ligand imaging markers/tracers of PET and cerebrospinal fluid (CSF) markers have been utilized to target various steps of neurotransmitter signaling in the diagnosis of AD and other neurologic diseases (Wolk et al., 2012).

5.5 Conclusion

PET-CT has enormous worth in locoregional and distant staging in both primary and recurring cancers. PET-CT is a handy tool to provide necessary aspects such as axillary lymph involvement. Further, the sensitivity and specificity are impressive except in the breast where the detection of small lesions was suboptimal. However, further research in combinational usage of radio tracers, FDG dosage, and combinational therapies is imperative in attaining better results. With the current research, PET-CT has a foreseeable potential role as a screening tool in high-risk cancer patients. FDG PET is extremely advantageous and useful in many cancer stagings because of its ability in detecting unsuspected distant metastases, which are missed in the conventional screening. Recently, PET has been utilized with MRI to improve better quality and accuracy of cancer diagnosis. It is also useful in personalized therapeutic management of various solid cancers. Other than cancer, PET scan is also useful in clinical diagnosis and research of human diseases, including cardiac, neurologic, and psychiatric disease.

References

Agarwal, A., Marcus, C., Xiao, J., Nene, P., Kachnic, L.A., Subramaniam, R.M., 2014. FDG PET/CT in the management of colorectal and anal cancers. American Journal of Roentgenology 203 (5), 1109–1119.

Ahuja, V., Coleman, R.E., Herndon, J., Patz Jr., E.F, 1998. The prognostic significance of fluorodeoxyglucose positron emission tomography imaging for patients with nonsmall cell lung carcinoma. Cancer 83 (5), 918–924.

Ali, Taha, F, T., 2012. Usefulness of PET–CT in the assessment of suspected recurrent colorectal carcinoma. The Egyptian Journal of Radiology and Nuclear Medicine 43 (2), 129–137.

Amthauer, H., Denecke, T., Rau, B., Hildebrandt, B., Hünerbein, M., Ruf, J., et al., 2004. Response prediction by FDG-PET after neoadjuvant radiochemotherapy and combined regional hyperthermia of rectal cancer: Correlation with endorectal ultrasound and histopathology. European Journal of Nuclear Medicine and Molecular Imaging 31 (6), 811–819.

Antoch, G., Stattaus, J., Nemat, A.T., Marnitz, S., Beyer, T., Kuehl, H., et al., 2003. Non-small cell lung cancer: Dual-modality PET/CT in preoperative staging. Radiology 229 (2), 526–533.

Avallone, A., Aloj, L., Caracò, C., Delrio, P., Pecori, B., Tatangelo, F., et al., 2012. Early FDG PET response assessment of preoperative radiochemotherapy in locally advanced rectal cancer: Correlation with long-term outcome. European Journal of Nuclear Medicine and Molecular Imaging 39 (12), 1848–1857.

Bastarrika, G., García-Velloso, M.J., Lozano, M.D., Montes, U., Torre, W., Spiteri, N., et al., 2005. Early lung cancer detection using spiral computed tomography and positron emission tomography. American Journal of Respiratory Critical Care Medicine 171 (12), 1378–1383.

Bellon, J.R., Livingston, R.B., Eubank, W.B., Gralow, J.R., Ellis, G.K., Dunnwald, L.K., et al., 2004. Evaluation of the internal mammary lymph nodes by FDG-PET in locally advanced breast cancer (LABC). American Journal of Clinical Oncology 27 (4), 407–410.

Bender, H., Kirst, J., Palmedo, H., Schomburg, A., Wagner, U., Ruhlmann, J., et al., 1997. Value of 18fluoro-deoxyglucose positron emission tomography in the staging of recurrent breast carcinoma. Anticancer Research 17 (3b), 1687–1692.

Bengel F.M., Higuchi T., Javadi M.S., Lautamäki R., Jun 30, 2009. Cardiac positron emission tomography. Journal of the American College of Cardiology. 54(1):1-15. doi: 10.1016/j.jacc.2009.02.065. PMID: 19555834.

Bipat, S., van Leeuwen, M.S., Comans, E.F., Pijl, M.E., Bossuyt, P.M., Zwinderman, A.H., et al., 2005. Colorectal liver metastases: CT, MR imaging, and PET for diagnosis–meta-analysis. Radiology 237 (1), 123–131.

Bunyaviroch, T., Coleman, R.E., 2006. PET evaluation of lung cancer. Journal of Nuclear Medicine 47 (3), 451–469.

Chapman, S.E., Diener, J.M., Sasser, T.A., Correcher, C., González, A.J., Avermaete, T.V., et al., 2012. Dual tracer imaging of SPECT and PET probes in living mice using a sequential protocol. American Journal of Nuclear Medicine and Molecular Imaging 2 (4), 405–414.

Ciernik, I.F., Dizendorf, E., Baumert, B.G., Reiner, B., Burger, C., Davis, J.B., et al., 2003. Radiation treatment planning with an integrated positron emission and computer tomography (PET/CT): A feasibility study. International Journal of Radiation, Oncology, Biology, Physics 57 (3), 853–863.

Ciernik, I.F., Huser, M., Burger, C., Davis, J.B., Szekely, G., 2005. Automated functional image-guided radiation treatment planning for rectal cancer. International Journal of Radiation, Oncology, Biology, Physics 62 (3), 893–900.

Cook, G.J., Fogelman, I., 2000. The role of positron emission tomography in the management of bone metastases. Cancer 88 (12), 2927–2933 Suppl.

Cronin, C.G., Swords, R., Truong, M.T., Viswanathan, C., Rohren, E., Giles, F.J., et al., 2010. Clinical utility of PET/CT in lymphoma. American Journal of Roentgenology 194 (1), W91–W103.

Dighe, S., Purkayastha, S., Swift, I., Tekkis, P.P., Darzi, A., A'Hern, R., et al., 2010. Diagnostic precision of CT in local staging of colon cancers: A meta-analysis. Clinical Radiology 65 (9), 708–719.

Downey, R.J., Akhurst, T., Gonen, M., Vincent, A., Bains, M.S., Larson, S., et al., 2004. Preoperative F-18 fluorodeoxyglucose-positron emission tomography maximal standardized uptake value predicts survival after lung cancer resection. Journal of Clinical Oncology 22 (16), 3255–3260.

Ferlay, J., Soerjomataram, I., Dikshit, R., Eser, S., Mathers, C., Rebelo, M., et al., 2015. Cancer incidence and mortality worldwide: Sources, methods and major patterns in GLOBOCAN 2012. International Journal of Cancer 136 (5), e359–e386.

Fletcher, J.W., Djulbegovic, B., Soares, H.P., Siegel, B.A., Lowe, V.J., Lyman, G.H., et al., 2008. Recommendations on the use of 18F-FDG PET in oncology. Journal of Nuclear Medicine 49 (3), 480–508.

Fuster, D., Duch, J., Paredes, P., Velasco, M., Muñoz, M., Santamaría, G., et al., 2008. Preoperative staging of large primary breast cancer with [18F]fluorodeoxyglucose positron emission tomography/computed tomography compared with conventional imaging procedures. Journal of Clinical Oncology 26 (29), 4746–4751.

Galldiks, N., Langen, K.J., Albert, N.L., Chamberlain, M., Soffietti, R., Kim, M.M., et al., 2019a. PET imaging in patients with brain metastasis-report of the RANO/PET group. Journal of Neuro-Oncology 21 (5), 585–595.

Galldiks, N., Lohmann, P., Albert, N.L., Tonn, J.C., Langen, K.J., 2019b. Current status of PET imaging in neuro-oncology. Neuro-Oncology Advances 1 (1) vdz010.

Garibotto, V., Heinzer, S., Vulliemoz, S., Guignard, R., Wissmeyer, M., Seeck, M., et al., 2013. Clinical applications of hybrid PET/MRI in neuroimaging. Clinical Nuclear Medicine 38 (1), e13–e18.

Georgakopoulos, A., Pianou, N., Kelekis, N., Chatziioannou, S., 2013. Impact of 18F-FDG PET/CT on therapeutic decisions in patients with colorectal cancer and liver metastases. Clinical Imaging 37 (3), 536–541.

Hetta, W., Niazi, G., Abdelbary, M.H., 2020. Accuracy of 18F-FDG PET/CT in monitoring therapeutic response and detection of loco-regional recurrence and metastatic deposits of colorectal cancer in comparison to CT. Egyptian Journal of Radiology and Nuclear Medicine 51 (1), 37.

Hochhegger, B., Alves, G.R., Irion, K.L., Fritscher, C.C., Fritscher, L.G., Concatto, N.H., et al., 2015. PET/CT imaging in lung cancer: Indications and findings. Jornal Brasileiro de Pneumologia 41 (3), 264–274.

Huang, B., Law, M.W., Khong, P.L., 2009. Whole-body PET/CT scanning: Estimation of radiation dose and cancer risk. Radiology 251 (1), 166–174.

Huang, S.W., Hsu, C.M., Jeng, W.J., Yen, T.C., Su, M.Y., Chiu, C.T., 2013. A comparison of positron emission tomography and colonoscopy for the detection of advanced colorectal neoplasms in subjects undergoing a health check-up. PLoS One 8 (7), e69111.

Huebner, R.H., Park, K.C., Shepherd, J.E., Schwimmer, J., Czernin, J., Phelps, M.E., et al., 2000. A meta-analysis of the literature for whole-body FDG PET detection of recurrent colorectal cancer. Journal of Nuclear Medicine 41 (7), 1177–1189.

Iqbal S., Iqbal K., Arif F., Shaukat A., Khanum A., 2014. Potential lung nodules identification for characterization by variable multistep threshold and shape indices from CT images. Computational and Mathematical Methods in Medicine 2014, 241647. doi: 10.1155/2014/241647. Epub 2014 Nov 25. PMID: 25506388; PMCID: PMC4260430.

Israel, O., Kuten, A., 2007. Early detection of cancer recurrence: 18F-FDG PET/CT can make a difference in diagnosis and patient care. Journal of Nuclear Medicine 48, S128–S135.

Kamel, E.M., Zwahlen, D., Wyss, M.T., Stumpe, K.D., von Schulthess, G.K., Steinert, H.C., 2003. Whole-body (18)F-FDG PET improves the management of patients with small cell lung cancer. Journal of Nuclear Medicine 44 (12), 1911–1917.

Kinkel, K., Lu, Y., Both, M., Warren, R.S., Thoeni, R.F., 2002. Detection of hepatic metastases from cancers of the gastrointestinal tract by using noninvasive imaging methods (US, CT, MR imaging, PET): A meta-analysis. Radiology 224 (3), 748–756.

Klein S., Staring M., Murphy K., Viergever M.A., Pluim J.P., Jan, 2010. elastix: a toolbox for intensity-based medical image registration. IEEE Transactions on Medical Imaging 29(1), 196-205. doi: 10.1109/TMI.2009.2035616. Epub 2009 Nov 17. PMID: 19923044.

Kochhar, R., Liong, S., Manoharan, P., 2010. The role of FDG PET/CT in patients with colorectal cancer metastases. Cancer Biomarkers 7 (4), 235–248.

Koolen, B.B., Vogel, W.V., Vrancken Peeters, M.J., Loo, C.E., Rutgers, E.J., Valdés Olmos, R.A., 2012. Molecular imaging in breast cancer: From whole-body PET/CT to dedicated breast PET. Journal of Oncology, 2012438647.

Kuipers, E.J., Rösch, T., Bretthauer, M., 2013. Colorectal cancer screening–optimizing current strategies and new directions. Nature Reviews Clinical Oncology 10 (3), 130–142.

Kunawudhi, A., Sereeborwornthanasak, K., Promteangtrong, C., Siripongpreeda, B., Vanprom, S., Chotipanich, C., 2016. Value of FDG PET/contrast-enhanced CT in initial staging of colorectal cancer—comparison with contrast-enhanced CT. Asian Pacific Journal of Cancer Prevention 17 (8), 4071–4075.

Lardinois, D., Weder, W., Hany, T.F., Kamel, E.M., Korom, S., Seifert, B., et al., 2003. Staging of non-small-cell lung cancer with integrated positron-emission tomography and computed tomography. New England Journal of Medicine 348 (25), 2500–2507.

Maeda, C., Endo, S., Mori, Y., Mukai, S., Hidaka, E., Ishida, F., et al., 2019. The ability of positron emission tomography/computed tomography to detect synchronous colonic cancers in patients with obstructive colorectal cancer. Molecular and Clinical Oncology 10 (4), 425–429.

Mahner, S., Schirrmacher, S., Brenner, W., Jenicke, L., Habermann, C.R., Avril, N., et al., 2008. Comparison between positron emission tomography using 2-[fluorine-18]fluoro-2-deoxy-D-glucose, conventional imaging and computed tomography for staging of breast cancer. Annals of Oncology 19 (7), 1249–1254.

Massoud T.F., Gambhir S.S., Mar 1, 2003. Molecular imaging in living subjects: Seeing fundamental biological processes in a new light. Genes and Development 17(5), 545-580. doi: 10.1101/gad.1047403. PMID: 12629038.

Moon, D.H., Maddahi, J., Silverman, D.H., Glaspy, J.A., Phelps, M.E., Hoh, C.K., 1998. Accuracy of whole-body fluorine-18-FDG PET for the detection of recurrent or metastatic breast carcinoma. Journal of Nuclear Medicine 39 (3), 431–435.

Murcia Duréndez, M.J., Frutos Esteban, L., Luján, J., Frutos, M.D., Valero, G., Navarro Fernández, J.L., et al., 2013. The value of 18F-FDG PET/CT for assessing the response to neoadjuvant therapy in locally advanced rectal cancer. European Journal of Nuclear Medicine and Molecular Imaging 40 (1), 91–97.

Niekel, M.C., Bipat, S., Stoker, J., 2010. Diagnostic imaging of colorectal liver metastases with CT, MR imaging, FDG PET, and/or FDG PET/CT: A meta-analysis of prospective studies including patients who have not previously undergone treatment. Radiology 257 (3), 674–684.

Nomori, H., Watanabe, K., Ohtsuka, T., Naruke, T., Suemasu, K., Uno, K., 2004. Evaluation of F-18 fluorodeoxyglucose (FDG) PET scanning for pulmonary nodules less than 3 cm in diameter, with special reference to the CT images. Lung Cancer 45 (1), 19–27.

O'Connor, O.J., McDermott, S., Slattery, J., Sahani, D., Blake, M.A., 2011. The use of PET-CT in the assessment of patients with colorectal carcinoma. International Journal of Surgical Oncology, 2011846512.

Patel, D.A., Chang, S.T., Goodman, K.A., Quon, A., Thorndyke, B., Gambhir, S.S., et al., 2007. Impact of integrated PET/CT on variability of target volume delineation in rectal cancer. Technology in Cancer Research and Treatment 6 (1), 31–36.

Rosé, C., Dose, J., Avril, N., 2002. Positron emission tomography for the diagnosis of breast cancer. Nuclear Medicine Communications 23 (7), 613–618.

Rottenburger, C., Hentschel, M., Kelly, T., Trippel, M., Brink, I., Reithmeier, T., et al., 2011. Comparison of C-11 methionine and C-11 choline for PET imaging of brain metastases: A prospective pilot study. Clinical Nuclear Medicine 36 (8), 639–642.

Sattler, B., Lee, J.A., Lonsdale, M., Coche, E., 2010. PET/CT (and CT) instrumentation, image reconstruction and data transfer for radiotherapy planning. Radiotherapy and Oncology 96 (3), 288–297.

Selzner, M., Hany, T.F., Wildbrett, P., McCormack, L., Kadry, Z., Clavien, P.A., 2004. Does the novel PET/CT imaging modality impact on the treatment of patients with metastatic colorectal cancer of the liver? Annals of Surgery 240 (6), 1027–1034 discussion 1035–1036.

Smith, I.C., Welch, A.E., Hutcheon, A.W., Miller, I.D., Payne, S., Chilcott, F., et al., 2000. Positron emission tomography using [(18)F]-fluorodeoxy-D-glucose to predict the pathologic response of breast cancer to primary chemotherapy. Journal of Clinical Oncology 18 (8), 1676–1688.

Spanoudaki, V., Levin, C.S., 2010. Photo-detectors for time of flight positron emission tomography (ToF-PET). Sensors (Basel) 10 (11), 10484–10505.

Strasberg, S.M., Dehdashti, F., Siegel, B.A., Drebin, J.A., Linehan, D., 2001. Survival of patients evaluated by FDG-PET before hepatic resection for metastatic colorectal carcinoma: A prospective database study. Annals of Surgery 233 (3), 293–299.

Tatsumi, M., Cohade, C., Mourtzikos, K.A., Fishman, E.K., Wahl, R.L., 2006. Initial experience with FDG-PET/CT in the evaluation of breast cancer. European Journal of Nuclear Medicine and Molecular Imaging 33 (3), 254–262.

Varrone, A., Asenbaum, S., Vander Borght, T., Booij, J., Nobili, F., Någren, K., et al., 2009. EANM procedure guidelines for PET brain imaging using [18F]FDG, version 2. European Journal of Nuclear Medicine and Molecular Imaging 36 (12), 2103–2110.

Welling M.M., Duijvestein M., Signore A., van der Weerd L., Jun, 2011. In vivo biodistribution of stem cells using molecular nuclear medicine imaging. Journal of Cellular Physiology 226(6), 1444-1452. doi: 10.1002/jcp.22539. PMID: 21413018.

Wolk, D.A., Zhang, Z., Boudhar, S., Clark, C.M., Pontecorvo, M.J., Arnold, S.E., 2012. Amyloid imaging in Alzheimer's disease: Comparison of florbetapir and Pittsburgh compound-B positron emission tomography. Journal of Neurology, Neurosurgery, Psychiatry 83 (9), 923–926.

Yang, S.K., Cho, N., Moon, W.K., 2007. The role of PET/CT for evaluating breast cancer. Korean Journal of Radiology 8 (5), 429–437.

Zhao, C., Zhang, Y., Wang, J., 2014. A meta-analysis on the diagnostic performance of (18) F-FDG and (11)C-methionine PET for differentiating brain tumors. American Journal of Neuroradiology 35 (6), 1058–1065.

Revolutionizing medical diagnosis with SPECT imaging: Clinical applications of a nuclear imaging technology

Mena Asha Krishnan[a]**, Amulya Cherukumudi**[b]**, Sibi Oommen**[c]**, Sumeet Suresh Malapure**[d]**, Venkatesh Chelvam**[a,e]

[a]*Department of Biosciences and Biomedical Engineering, Indian Institute of Technology Indore, Khandwa Road, Simrol, Indore, Madhya Pradesh*
[b]*Department of General Surgery, The Bangalore Hospital, Bangalore*
[c]*Dept. of Nuclear Medicine, Manipal College of Health Professions, Manipal Academy of Higher Education, Manipal, Udupi, Karnataka*
[d]*Nuclear Medicine division, Kasturba Medical College, Manipal Academy of Higher Education, Manipal, Udupi, Karnataka*
[e]*Department of Chemistry, Indian Institute of Technology Indore, Khandwa Road,Simrol, Indore, Madhya Pradesh*

6.1 Introduction

6.1.1 Historical origin and development of single photon emission tomography

Nuclear medicine, a branch of medicine, uses radiotracers or radiopharmaceutics for organ function assessment as well as treatment. The use of radioisotopes for nuclear diagnosis began in early 1900. Notable contributions include:

Radioisotope	Year	Application
Phosphorus-32	1936	Treatment of leukemia
Technetium-99m	1937	Imaging various organs
Iodine-131	1955	Liver imaging
Xenon-133	1957	Lung imaging
Radiomercury (Chlormerodrin)	1960	Kidney imaging

To trace and visualize the biochemical process using various radiopharmaceutics, imaging modality became an integral part of nuclear medicine. Contributions by Benedict Cassen in 1952 in developing the first imaging modality rectilinear scanner

Biomedical Imaging Instrumentation. **DOI: https://doi.org/10.1016/B978-0-323-85650-8.00010-3**

and Hal Oscar Anger in 1957 for the Anger camera were remarkable and the base of other nuclear or molecular imaging modalities.

Other imaging milestones include:

a. Gamma camera in 1963 by Anger et al.
b. General-purpose single photon emission computed tomography (SPECT) camera in 1976 by John Keyes
c. First dedicated head SPECT camera in 1976 by Ronald Jaszczak
d. The first successful SPECT imaging of a neuroreceptor in humans in 1983 by William Eckelman and Richard Reba
e. First PET/CT prototype in 1998 by D.W.Townsend

6.2 **Components and working of SPECT imaging device**

For SPECT imaging, gamma camera detectors rotate around the organ of interest of a patient and acquire a series of images at a definite angle (Fig. 6.1). Detectors of SPECT imaging devices could be either single, dual, or triple headed. In addition to multiple slices, SPECT imaging aids in removing the overlying and underlying structures from the region of interest and enhances the contrast of the image.

Broadly, components of a gamma camera and SPECT imaging device consist of a detector (collimator, detector, photomultiplier tube, and preamplifier), processing (position logic circuit, amplifier, correction circuit, pulse height analyzer), and display (monitor) units (Cherry et al., 2003). Collimator, a sheet with holes/septa made of high atomic number and density material, helps in discriminating photons arising

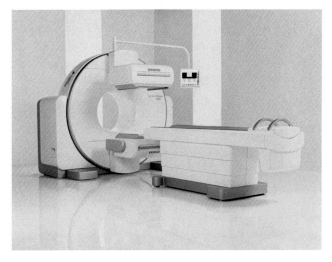

FIG. 6.1

SIEMENS SPECT/CT scanner Image Courtesy Siemens Healthcare GmbH.

from the tissue as good or bad based on the direction it travels toward the scintillation crystal. The choice of material includes lead, tungsten, gold, depleted uranium, among others. Based on energy, resolution, sensitivity, and design, collimators are classified into various categories. Parallel hole, pinhole, converging, and diverging collimators are classifications based on design. Energy-sensitive scintillation crystal/detector converts the incident photons passing through the collimator to its corresponding light photons, sodium iodide doped with thallium being the most common detector. Photomultiplier tube (PMT) converts the scintillation photons to a detectable electronic signal. It consists of a photocathode and an anode. The photocathode, a photoemissive surface, receives the scintillation light from the crystal and emits photoelectrons equivalent to initial incident radiation energy. The anode collects the amplified electrons through the multistage dynodes and outputs the electron current to an external circuit. Preamplifier, the next component, matches the impedance between the PMT and the main amplifier, in addition, to signal amplification and shaping. In the gamma camera, the position logic circuit helps an event to be localized in the appropriate image matrix (Prekeges, 2011). The amplifier further amplifies and shapes the signal, thereby improving the signal-to-noise ratio. To discriminate between good and bad signals based on energy, a pulse height analyzer (PHA) measures the amplitude of each pulse and compares them to preset values stored within it. There are two types of PHAs: single channel and multichannel. Good signals with all corrections are used for image creation and display. Images are nothing but the distribution of counts in matrix (Fig. 6.2).

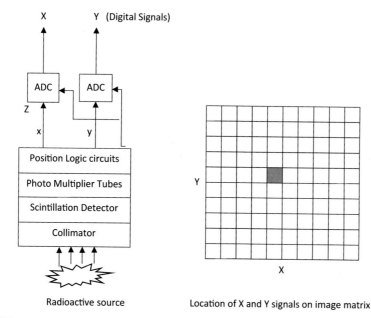

FIG. 6.2

Basic working principle of the gamma camera and image creation.

The gamma imaging recorded using a gamma camera is used in many diseases across the medical fields, including neurology, cardiology, and oncology. However, as it is two dimensional (2D) imaging, it has some inherent limitations. There was a need to acquire and see the images in the three-dimensional (3D) mode, which will help to localize the exact site of the disease beneath the anatomic complexity of the human body. The work started in 1963 and came to clinical picture a decade later, imaging evolution taking place from a rotating chair in front of the gamma camera and later rotating camera itself (Bowley et al., 1973; Kuhl & Edwards, 1963, 1970; Kuhl et al., 1966). Now we have full-fledged hybrid SPECT/computed tomography (CT) scanners as described later in this chapter.

In SPECT 2D, projections generated from multiple angles undergo reconstruction to generate object activity distribution in a 3D image matrix (Groch & Erwin, 2000; Prekeges, 2011). The various reconstruction methods include simple back-projection, direct Fourier transform (FT) reconstruction, filtered back-projection, and iterative reconstruction. In a simple back-projection technique, projections are back-projected on the image grid along the angle of acquisition. The resultant image would be blurred with star or streak artifact (Fig. 6.3). In direct FT reconstruction technique employing projection slice theorem, one-dimensional (1D) projection profiles in object space are taken into k space or Fourier space to determine the 2D k space data. Upon taking the inverse FT, the image can be computed. Though computationally intensive, an exact representation of objects is possible with noise-free data. Filtered back-projection techniques combine back-projection techniques with projection slice theorem using appropriate filters or kernels (Bruyant, 2002; Lyra & Ploussi, 2011)-Ramp, Butterworth, Shepp-Logan, Hann, Hamming, Parzen filter, etc., to name a few. Nyquist frequency, the highest spatial frequency the system can

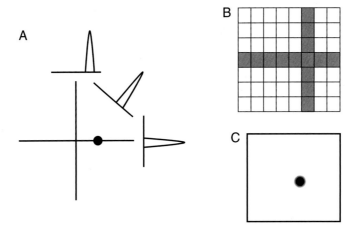

FIG. 6.3

(A) Acquiring projection profiles from various angles. (B) Back projection across the matrix at an angle of acquisition. (C) Blurred smoothened image.

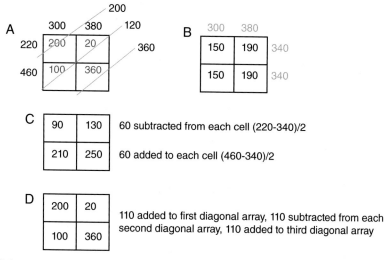

FIG. 6.4

Illustration for iterative reconstruction (A) Projections with actual measurements (B) First initial guess with a comparison based on vertical projections (C) Corrected matrix on comparison with horizontal projections (D) Corrected matrix on comparison with diagonal projections.

reproduce, needs to be considered while applying filters to avoid artifacts. In the iterative reconstruction technique, an initial guess and estimated 3D image matrix is created. Measured 2D projections are compared with estimated 2D, differences between them are used to create a 3D error matrix, which aids in updating the initial estimated 3D image matrix (Fig. 6.4). Among iterative reconstruction techniques incorporating various statistical and technical factors, maximum likelihood expectation maximization (MLEM) and ordered subsets expectation maximization (OSEM) became popular (Hutton, 2011). FLASH 3D and spline reconstruction are other techniques available. For accurate quantification and enhanced image quality, the data acquired from SPECT device need to be corrected for attenuation and scatter prior to reconstruction. CT aids in attenuation correction in addition to images with high resolution and contrast. All these corrections can be incorporated into the iterative reconstruction technique. The display of transverse, coronal, and sagittal slices includes the final step in SPECT. 3D display techniques include surface rendering, volume rendering, and maximum intensity projection (MIP) images. The factors that influence the quality of image includes choice and quality of radiopharmaceutic, pathology of the disease, choice of the collimator, energy window, matrix size, zoom, detector orbit, mode of acquisition, number of views, time per stop, noise, choice, and application of filters, patient motion, presence of implants, contrast, attenuation, scatter, and partial volume effect. In addition to gamma camera quality control tests,

SPECT imaging devices do have additional tests that need to give good results for excellent image quality, especially uniformity and center of rotation.

6.3 SPECT applications

6.3.1 SPECT/CT for benign conditions of the bone

Some of the benign indications for bone SPECT are to diagnose osteoid osteoma and the condylar hyperplasia of the mandible. Osteoid osteoma usually appears as a hot focal spot on planar images; however, in some patients, the location of the osteoma is at places where it can be easily mistaken for physiologic uptake like the proximal end of the femur or in the posterior elements of the vertebrae where it can be mistaken for degenerative change. Here SPECT/CT helps localize the lesion and exclude other differentials. Coming to condylar hyperplasia, any changes appearing in the condyles of the mandible, which is hidden behind the temporomandibular joint, is difficult to appreciate on planar imaging. SPECT/CT helps us in identifying these subtle changes and also quantify the active lesions as opposed to the planar images (López & Corral, 2016) (Fig. 6.5). Similarly, with diabetic foot and osteomyelitis, where the involved bone usually shows diffusely increased radiotracer uptake, SPECT/CT identifies if the bone is involved or not or whether the part of the affected bone is viable or not.

6.3.2 SPECT/CT in neuroendocrine neoplasms

A. ^{123}I-meta-iodobenzylguanidine (MIBG) scintigraphy is used to image and treat neuroendocrine tumors such as pheochromocytoma-paranganglioma, neuroblastoma, medullary carcinoma of the thyroid, among others. The anatomic locations of these lesions, particularly the adrenals, are difficult to diagnose with planar imaging. SPECT/CT imaging being more sensitive with better spatial resolutions helps clinch the diagnosis. Also, with the CT part we can differentiate accurately between the physiologic uptake of the tracer versus disease involvement (Castellani et al., 2008).

B. ^{131}I-Iodide SPECT/CT in thyroid cancer is used for the evaluation of patients with differentiated thyroid cancer for residual disease after thyroidectomy. This provides a benefit over the whole body ^{131}I scan, as that can miss poorly/nonfunctioning lesions. Moreover, a greater number of lesions are noted with SPECT. Hybrid SPECT/CT scan provides accurate localization and size estimation of radioiodine foci and iodine-avid metastases (Bhattacharya et al., 2010). This helps to adequately plan the high-dose radioiodine ablative therapy in such patients.

C. ^{99m}Tc-Sestamibi for parathyroid imaging: Currently, minimally invasive parathyroid surgery is the treatment of choice for parathyroid adenomas. Hence it becomes necessary to identify the location of the adenoma prior to the surgery. These adenomas, due to their anatomic location, are difficult to

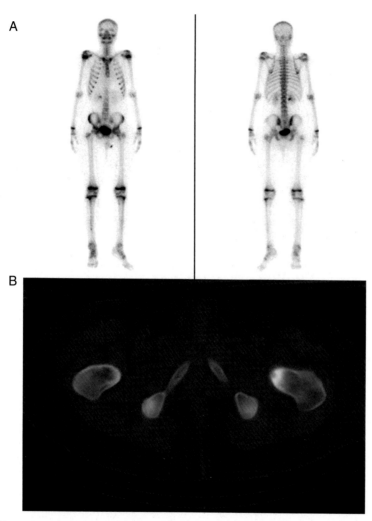

FIG. 6.5

(A) A 16 yr old boy with a history of left hip pain was referred for a bone scan. The planar whole-body bone scan of the boy shows no definite abnormal radiotracer uptake in the hip bones. (B) SPECT/CT of the hip reveals subtle cortical thickening with central lucency in the neck of the left femur, which was consistent with osteoid osteoma.

identify on ultrasound and on planar Sestamibi scans. Sestamibi SPECT/CT is currently the gold standard imaging modality for diagnosis and identification of parathyroid adenomas (Patel, 2010). Even subcentimetric adenomas are identified using SPECT (Fig. 6.6); however, parathyroid hyperplasia is usually missed on SPECT.

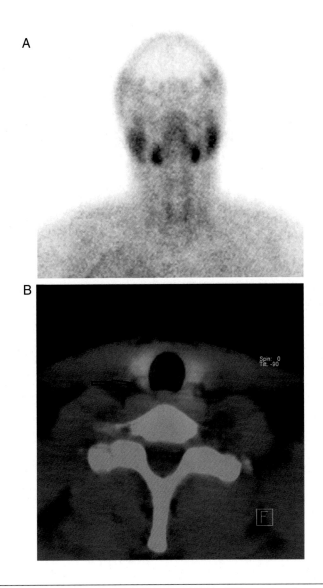

FIG. 6.6

(A) A 62 yr old female patient with mildly increased PTH of 81 pg/ml and a normal ultrasound was referred for ^{99m}Tc-Sestamibi parathyroid scan. A planar static image of the patient did not show any abnormal radiotracer uptake in the parathyroid region. (B) SPECT/CT reveals a subcentimetric soft tissue density lesion showing mildly increased radiotracer uptake posterior to the lower pole of the right lobe of the thyroid. The patient underwent minimally invasive parathyroid surgery, and the histopathology confirmed it as parathyroid adenoma.

6.3.3 **Radionuclide bone imaging for staging malignant metastases**

a. **Risks of skeletal metastasis**

The human skeleton is known to be one of the most common sites for malignant tumor metastasis, found to occur in 30% to 70% of patients (Even-Sapir, 2005). The treatment of solid tumors usually involves surgery, chemotherapy, and radiotherapy, or a combination of all three. The treatment success and high survival rate depend on the extent of localization of the tumor to the region of origin. A much poorer outcome is associated with the metastatic spread of malignancy, which invades lymph nodes, distant organs, and attacks the skeleton. Cancers of the lung (D'Antonio et al., 2014), breast (Manders et al., 2006), and prostate (Carlin & Andriole, 2000) are the major sources of metastasis, followed by bladder (Zhang, 2018), thyroid (Durante et al., 2006), and uterine malignancies (Uccella et al., 2013). Some of the comorbidities patients experience with bone metastasis include immense joint pain, compression of the spinal cord and nerve endings, hypercalcaemia, and limited mobility along with pathologic fractures (Cuccurullo et al., 2013).

b. **Mechanism of the metastatic process in the skeletal system**

In general, bone is always in a constant state of dynamic remodeling, maintaining a balance between resorption of aged bone matter and formation of new bone. This equilibrium is essential for the maintenance of normal skeletal structure and function. The two main cell types involved in this remodeling are osteoclasts (bone resorption activity) and osteoblasts (bone formation activity) (Morgan-Parkes, 1995). In the case of malignant growths, however, metastatic lesions travel from breast, lung, prostate, among others, and begin to enlarge within the marrow. This causes reactive changes in the osteoclastic and osteoblastic equilibrium resulting in three types of bone metastasis such as lytic, sclerotic (blastic), and mixed. Among these, rapidly proliferating, aggressive metastasis are lytic lesions, whereas a less aggressive form is seen in the sclerotic lesions (Hamaoka et al., 2004).

It should be noted that most patients with malignant bone lesions are asymptomatic, and the malignancies are detected only by routine screening. Symptoms manifest only in the case of highly enlarged lesions, which causes destruction of the bone, fractures, and in some cases spinal cord or nerve ending compression. Currently, diagnosis and management are performed using integrated systems for SPECT or SPECT/CT hybrid imaging in routine practice.

c. **SPECT imaging for detecting bone metastasis**

The major added advantage of SPECT over planar gamma imaging in skeletal scintigraphy apart from being more sensitive is to differentiate between benign and malignant bone lesions (Rager et al., 2017). As bone tracers like ^{99m}Tc-MDP are very nonspecific in nature, it becomes necessary to accurately characterize the lesion, which in turn will change the stage of the disease and hence treatment (Rager et al., 2017) (Fig. 6.7). The first hybrid system combining the SPECT gamma camera and a CT scanner in a single framework became available approximately 12 years ago (Seo et al., 2008). Since then, this

A

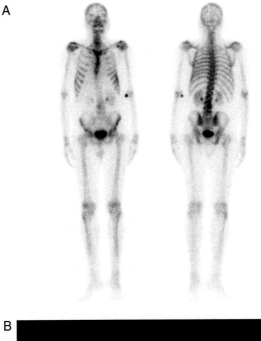

B

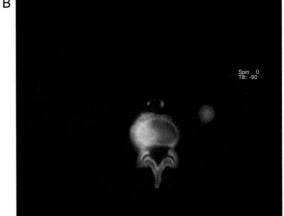

FIG. 6.7

(A) Planar whole-body bone scan of a 55yr old female patient with breast cancer showing increased radiotracer uptake in the lumbar vertebrae. (B) SPECT/CT reveals osteophytic changes in the lumbar vertebrae showing increased radiotracer uptake, which are degenerative in nature.

nuclear imaging modality has allowed accurate detection of bone metastasis in a timely and cost-effective routine, localization of SPECT foci of abnormal tracer uptake, diagnosis of morphologic abnormalities, and attenuation correction mechanism for more uniform images (Beyer & Veit-Haibach, 2014).

d. **99mTc-methylene diphosphonate (^{99m}Tc-MDP) tracer for skeletal scintigraphy**

For more than half a century now, ^{99m}Tc radioisotope has been regarded as a standard radiotracer for medical diagnosis. ^{99m}Tc radioisotope is produced in high yield, and purity from a ^{99}molybdenum (^{99}Mo)/^{99m}technetium (^{99m}Tc) generator and its ease of production onsite, favorable half-life, facile radiolabeling chemistry, and image quality have made it highly favorable in bone imaging (Herbert et al., 1965). A radionuclide bone scan using ^{99m}Tc-radiolabeled bisphosphonates is hence the preferred method to detect metastatic progression in the skeleton. The principle used here is that all bisphosphonates show a high affinity to the bone. The coordination of the phosphate group in the bisphosphonate with calcium present in the hydroxyapatite of bone results in the accumulation of the radiolabeled complexes to active sites in bone structure (Lin, 1996). This accumulation is imaged using a planar gamma camera, which allows whole-body visualization of the spatial distribution of the ^{99m}Tc-tracer via a 140 keV γ-ray produced upon radioactive decay. The absorption and scatter patterns of the photons are collected in both anterior and posterior planar scans, which are finally read by a certified radiologist who identifies the malignant lesions. In contrast to CT, bone scintigraphy readings can detect malignant growth 18 months earlier in patients. In the Western world, medronic acid is a commonly used phosphate to form ^{99m}Tc-methylene diphosphonate (^{99m}Tc-MDP) tracer due to its relatively high bone uptake, fast clearance from tissues, and stability in blood plasma (Einhorn, 1986).

6.4 Targeted imaging of prostate cancer with radioligands

a. **The necessity for targeted imaging of prostate cancer**

World statistics have shown that prostate cancer is the second most common cause of death in the Western world and the most commonly occurring solid cancer in men (Siegel, 2019). Multiple therapeutic courses such as chemotherapy, androgen therapy, radiotherapy, and surgery, or a combination of these strategies, are available based on the risk of disease, which is determined by the prostate-specific antigen (PSA) serum level, Gleason score, and imaging results (Litwin & Tan, 2017). Conventional diagnostic options include a digital rectal exam (DRE) test, ultrasonography, CT, magnetic resonance imaging (MRI), and whole-body scan. However, these modalities can perceive only morphologic and anatomic changes in organs and fail to detect abnormalities arising due to pathologic and inflammatory pathways (Hricak et al., 2007). It is for this reason that molecular imaging of prostate cancer is performed, where targeted radiolabeled tracers are injected into the patient and then monitored for uptake in abnormal or malignant tissues. It is known that cancer cells overexpress certain proteins called biomarkers during diseased conditions that facilitate accurate and sensitive detection of cancer. Hence radiotracers chelated to ligands or small molecules that specifically bind to these biomarkers are targeted to image prostate cancer, ensuring that the targeted radiotracer detects only cancerous cells and not normal tissues.

b. **Prostate-specific membrane antigen for detection of prostate cancer**

The discovery of the prostate-specific membrane antigen (PSMA) biomarker, expressed 100- to 1000-fold on cancerous cells when compared to normal cells, is the most popular biomarker among the other 91 biomarkers known, whose expression on cells also correlates with the extent of disease progression (Tricoli et al., 2004). PSMA is a type II membrane glycoprotein that consists of an extracellular peptide domain of 44 to 750 amino residues, a hydrophobic transmembrane region, and a short NH_2 terminal cytoplasmic domain. PSMA possesses an enzymatic ability to cleave the terminal glutamate moiety from neuropeptide N-acetyl-aspartyl glutamate (NAAG), and hence it is also called glutamate carboxypeptidase II or folate hydrolase (Rajasekaran et al., 2005). Recent reports have also found PSMA to be expressed in the endothelial cells of tumor-vasculature and absent in healthy blood vessels (Chang et al., 1999). Collectively, these characteristics make PSMA an ideal target to deliver diagnostic agents to detect prostate cancer cells in the system.

c. **SPECT radionuclear-targeted imaging of PSMA**

For improved imaging and better detection of prostate cancer, many small-molecule inhibitors have been designed to target the extracellular domain of PSMA. For this, a number of inhibitor scaffolds were developed, of which glutamate-urea-lysine and glutamate-urea-glutamate scaffold have attained the highest popularity due to their high affinity to PSMA receptor, ease of synthesis, and conjugation to linkers for attachment of radionuclides for prostate cancer detection (Sengupta et al., 2019).

^{99m}Tc has become one of the main radioisotopes for nuclear imaging using SPECT technology, and it is used more than 85% of the time annually, in some chemical form or the other in diagnostic scans in hospitals. ^{99m}Tc radiopharmaceutic emits 140-keV γ-ray with 89% abundance and is easily available from commercial generators, making it the preferred radioisotope for imaging using SPECT gamma camera. These characteristics have led to the development of imaging probes such as ^{99m}Tc-MIP-1404 with a high affinity of $k_d = 1.07 \pm 0.89$ nM for PSMA expressed on prostate cancer cells. MIP-1404 inhibitor core is based on the glutamate-urea-glutamate inhibitor structure and contains a bis-imidazole chelator capable of complexing Tc(CO)$^{3+}$. ^{99m}Tc-MIP-1404 has proved to be successful in phase I clinical studies, showing specific targeting of metastatic prostate cancer (Hillier et al., 2013).

Another high-affinity inhibitor for optimal targeting of PSMA protein is the small molecule urea-based ligand, DUPA (2-[3-(1,3-dicarboxypropyl) ureido] pentanedioic acid). This symmetric urea monomer derived from L-glutamic acid has shown binding affinity, $K_i = 8$ nM, to the PSMA protein (Kularatne et al., 2009). A radiopharmaceutic developed by the Indiana University Medical School in the United States is the first ^{99m}Tc-chelated small molecule nuclear imaging agent to enter clinical trials. Preliminary investigations of ^{99m}Tc-DUPA conjugate in prostate cancer xenografts implanted in mice showed excellent uptake of 12.4% ID/g tumor at 4 hours postinjection, which in turn increases the signal-to-noise ratio and tumor retention time, thereby providing greater contrast between cancer and background tissues (Banerjee et al., 2008).

6.5 SPECT imaging using 67ga-citrate SPECT/CT in lymphoma

Hodgkin disease (HD) and non-Hodgkin lymphoma (NHL), a malignancy of the lymphocytes of the immune system, is a disease that forms one of the most treatable cancers, where more than 90% of the patients survive for 5 years or greater (Shenoy et al., 2011). However, initial staging and monitoring of therapeutic effectiveness are highly recommended for a cure. Current diagnostic modalities employed provide only morphologic information (CT and MRI). However, the past decade has found SPECT imaging using ^{67}Ga-citrate (^{67}Ga) radiopharmaceutic to be of immense importance in the evaluation of viable lymphoma. As ^{67}Ga-citrate can easily distinguish between viable residual tumor after treatment and nonmalignant fibrotic tissue, it is also widely used to monitor the response after treatment (Fuertes Manuel et al., 2006). ^{67}Ga-citrate SPECT scan in lymphoma has a number of advantages, including (a) accurate discrimination of normal physiologic uptake of tracer and uptake in the foci of lymphoma when located close by, (b) abnormal gallium uptake can be precisely defined to specific lymph nodes, indicating the origin of lymphoma, and (c) accurate detection of residual lymphoma lesions within a heterogeneous tissue mass. It should be noted that among all the modalities, ^{67}Ga-citrate SPECT scan provides functional information about the lymphoma (Palumbo et al., 2005).

6.6 SPECT imaging: nononcologic diseases

a. Myocardial perfusion imaging (MPI): The primary role of MPI is to check whether the coronary stenotic lesion is hemodynamically significant and to assess whether the viability of the infarcted myocardium is viable or not. In most cases, we need to see one of the walls of the myocardium, which is affected. The heart is a globular structure having rhythmic movement that throws specific challenges for imaging that cannot be answered by planar imaging. This is solved using "gated" SPECT, which gives us multiple views of all the walls of the heart and also a structural map called polar map to identify and quantify the amount of myocardium affected. SPECT/CT provides the added benefit of risk stratifications as well as guiding the interventional radiologist/cardiologist in the choice of the revascularization procedure. With the newer development of organ-specific SPECT cameras like dedicated cardiac SPECT, significant strides have been made in the research of measuring myocardial blood flow and coronary flow reserves, which is the future of functional myocardial imaging (Moshe et al., 2010).

b. SPECT imaging of brain perfusion: The brain provides a unique challenge to imaging. The homogeneity within the complex neuro structure makes it virtually impossible to locate the diseased site on a planar gamma imaging; and in the majority of the diseases, functional defects predate anatomic changes. SPECT imaging holds a significant mileage over other imaging modalities in identifying neurologic problems, particularly in movement disorders and epilepsy. In intractable epilepsy patients, surgery provides a curative treatment; however, it is of paramount importance to identify the epileptogenic foci. Perfusion ictal and interictal SPECT scans help us identify the focus, which can be fused with MRI aiding the neurosurgeon in planning surgical resection.In movement disorders, a very sensitive scan called DAT scan is used to identify the loss of dopaminergic neurons of the basal ganglia. As you can see, planar imaging does not provide any insights regarding the disease, whereas SPECT clearly shows the neuronal defect (Fig. 6.8).

c. Lung disorders: Pulmonary and perfusion (V/Q) imaging is used in determining the presence of pulmonary thromboembolism, as well as the extent of the same within each anatomic segment of the lung. V/Q SPECT/CT has higher sensitivity over planar 2D scintigraphy, as it eliminates the anatomic overlap that is seen in the 2D planar scans. It is also used in quantification of lung function and assessment of lung tissue during asthma, emphysema, and interstitial lung disease. This can be further combined with pulmonary angiography to improve the sensitivity and specificity of V/Q SPECT (Roach et al., 2006).

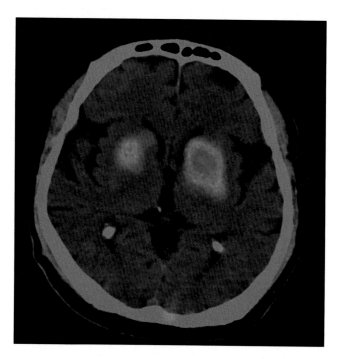

FIG. 6.8

A 62yr old male patient with left upper limb tremors was sent for DAT imaging after the MRI did not reveal any abnormalities. TRODAT SPECT/CT of the patient reveals a significant decrease in radiotracer uptake in the right putamen consistent with early Parkinson's disease.

6.7 Conclusion

Nuclear medicine uses radioactive probes called radiotracers to diagnose various diseases and has revolutionized radiologic imaging since its inception. Of these, SPECT plays a pivotal role in providing information about the physiology and pathology of various ailments, helping in improving sensitivity and specificity. To obtain these images, the gamma camera rotates around the organ of interest and generates a series of images, which can be analyzed to isolate the defect. This is used across various systems, such as neurology, cardiology, and oncology. Initially, SPECT images were obtained in 2D, which had its own limitations. However, acquiring and generating 3D images can overcome the anatomic complexities of the person undergoing the investigation. Hybrid SPECT/CT can be used to evaluate benign conditions of the bone, skeletal metastases, neuroendocrine tumors, thyroid cancer, and localization of parathyroid for minimally invasive parathyroid surgery. There has been immense progress in the development of the radiotracers being used, which improves the sensitivity and specificity of the scans further. Several improvements have been made in the various components of the machine, the gamma camera, the image rendering process, and the radiotracers being used. Improvements in the scintillators and the photon transducers allow the devices to fabricate smaller and more compact systems due to the improvements in semiconductors. New clinical devices include high-count sensitivity cardiac SPECT systems that do not use conventional collimation and the introduction of diagnostic-quality hybrid SPECT/CT systems. Despite the progress with the existing reconstruction algorithms, newer versions have been made commercially available to reduce the SPECT acquisition time with better quality images. Preclinical small-animal SPECT systems have become a major focus in nuclear medicine. These systems have pushed SPECT limits into the submillimeter range, making them valuable molecular imaging tools capable of providing information unavailable from other modalities.

References

Banerjee, S.R., Foss, C.A., Castanares, M., Mease, R.C., Byun, Y., Fox, J.J., et al., 2008. Synthesis and evaluation of technetium-99m- and rhenium-labeled inhibitors of the prostate-specific membrane antigen (PSMA). J. Med. Chem. 51, 4504–4517. https://doi.org/10.1021/jm800111u.

Beyer, T., Veit-Haibach, P., 2014. State-of-the-art SPECT/CT: Technology, methodology and applications-defining a new role for an undervalued multimodality imaging technique. Eur. J. Nucl. Med. Mol. Imaging 41, 1–2. https://doi.org/10.1007/s00259-014-2696-8.

Bhattacharya, A., Venkataramarao, S.H., Bal, C.S., Mittal, B.R., 2010. Utility of iodine-131 hybrid SPECT-CT fusion imaging before high-dose radioiodine therapy in papillary thyroid carcinoma. Indian J. Nucl. Med. 25 (1), 29–31. https://doi:10.4103/0972-3919.63599.

Bowley, A.R., Taylor, C.G., Causer, D.A., Barber, D.C., Keyes, W.I., Undrill, P.E., et al., 1973. A radioisotope scanner for rectilinear, arc, transverse section and longitudinal section scanning: (ASS – the Aberdeen Section Scanner). Br. J. Radiol 46, 262–271.

Bruyant, P.P., 2002. Analytic and iterative reconstruction algorithms in SPECT. Journal of Nuclear Medicine 45, 1343–1358.

Carlin, B.I., Andriole, G.L., 2000. The natural history, skeletal complications, and management of bone metastases in patients with prostate carcinoma. Cancer 88, 2989–2994. https://doi.org/10.1002/1097-0142(20000615)88:12+<2989::aid-cncr14>3.3.co;2-h.

Castellani, M.R., Seregni, E., Maccauro, M., Chiesa, C., Aliberti, G., Orunesu, E., Bombardieri, E., 2008. MIBG for diagnosis and therapy of medullary thyroid carcinoma: Is there still a role? Q. J. Nucl. Med. Mol. Imaging 52, 430–440 PMID: 19088696.

Chang, S.S., Reuter, V.E., Heston, WDW, Bander, NH, Grauer, LS, Gaudin, PB., 1999. Five different anti-prostate-specific membrane antigen (PSMA) antibodies confirm PSMA expression in tumor-associated neovasculature. Cancer Res. 59, 3192–3198. https://doi.org/10.1006/geno.1995.0019.

Cherry, S.R., Sorenson, J.A., Phelps, M.E., 2003. Physics in Nuclear Medicine, 3rd edition. Saunders, Philadelphia, Pa, USA.

Cuccurullo, V., Lucio Cascini, G., Tamburrini, O., Rotondo, A., Mansi, L., 2013. Bone metastases radiopharmaceuticals: An overview. Curr. Radiopharm. 6, 41–47. https://doi.org/10.2174/1874471011306010007.

D'Antonio, C., Passaro, A., Gori, B., Del Signore, E., Migliorino, M.R., Ricciardi, S., et al., 2014. Bone and brain metastasis in lung cancer: Recent advances in therapeutic strategies. Ther Adv Med Oncol 6, 101–114. https://doi.org/10.1177/1758834014521110.

Durante, C., Haddy, N., Baudin, E., Leboulleux, S., Hartl, D., Travagli, J.P., et al., 2006. Long-term outcome of 444 patients with distant metastases from papillary and follicular thyroid carcinoma: Benefits and limits of radioiodine therapy. J. Clin. Endocrinol. Metab. 91, 2892–2899. https://doi.org/10.1210/jc.2005-2838.

Einhorn, T.A., Vigorita, V.J., Aaron, A., 1986. Localization of 99mTc methylene diphosphonate in bone using microautoradiography. J. Orthop. Res. Soc. 4, 180–187. https://doi.org/10.1002/jor.1100040206.

Even-Sapir, E., 2005. Imaging of malignant bone involvement by morphologic, scintigraphic, and hybrid modalities. J. Nucl. Med. 46, 1356–1367.

Fuertes Manuel, J., Estorch Cabrera, M., Camacho Martí, V., Flotats Giralt, A., Rodríguez-Revuelto, A.A., Hernández Fructuoso, M.A., et al., 2006. SPECT-CT 67Ga studies in lymphoma disease. Contribution to staging and follow-up]. Rev. Esp. Med. Nucl. 25, 242–249. https://doi.org/10.1157/13090657.

Groch, M.W., Erwin, W.D., 2000. SPECT in the year 2000: Basic principles. J. Nucl. Med. Tech. 28, 233–244.

Hamaoka, T., Madewell, J.E., Podoloff, D.A., Hortobagyi, G.N., Ueno, N.T., 2004. Bone imaging in metastatic breast cancer. J. Clin. Oncol. 22, 2942–2953. https://doi.org/10.1200/JCO.2004.08.181.

Herbert, R., Kulke, W., Shepherd, R.T., 1965. The use of 99mTc as a clinical tracer element. Postgrad Med. J. 41, 656–662. https://doi.org/10.1136/pgmj.41.481.656.

Hillier, S.M., Maresca, K.P., Lu, G., Merkin, R.D., Marquis, J.C., Zimmerman, C.N., et al., 2013. 99mTc-labeled small-molecule inhibitors of prostate-specific membrane antigen for molecular imaging of prostate cancer. J. Nucl. Med. 54, 1369–1376. https://doi.org/10.2967/jnumed.112.116624.

Hricak, H., Choyke, P.L., Eberhardt, S.C., Leibel, S.A., Scardino, P.T., 2007. Imaging prostate cancer: A multidisciplinary perspective. Radiology 243, 28–53. https://doi.org/10.1148/radiol.2431030580.

Hutton, B.F., 2011. Recent advances in iterative reconstruction for clinical SPECT/PET and CT. Acta Oncologica 50, 851–858.

Kuhl, D.E., Edwards, R.Q., 1963. Image separation radioisotope scanning. Radiology 80, 653–662.

Kuhl, D.E., Edwards, R.Q., 1970. The Mark III scanner: A compact device for multiple-view and section scanning of the brain. Radiology 96, 563–570.

Kuhl, D.E., Hale, J., Eaton, W.L., 1966. Transmission scanning: A useful adjunct to conventional emission scanning for accurately keying isotope deposition to radiographic anatomy. Radiology 87, 278–284.

Kularatne, S.A., Zhou, Z., Yang, J., Post, C.B., Low, P.S., 2009. Design, synthesis, and preclinical evaluation of prostate-specific membrane antigen targeted 99mTc-radioimaging agents. Mol. Pharm. 6, 790–800.

Lin, J.H., 1996. Bisphosphonates: A review of their pharmacokinetic properties. Bone 18, 75–85. https://doi.org/10.1016/8756-3282(95)00445-9.

Litwin, M.S., Tan, H-J., 2017. The diagnosis and treatment of prostate cancer: A review. JAMA 317, 2532–2542. https://doi.org/10.1001/jama.2017.7248.

López, B. DF., Corral, S. CM., 2016. Comparison of planar bone scintigraphy and single photon emission computed tomography for diagnosis of active condylar hyperplasia. J. Craniomaxillofac Surg. 44, 70–74.

Lyra, M., Ploussi, A., 2011. Filtering in SPECT Image Reconstruction. Int. J. Biomed. Imag. (2011), 14 pages Article ID 693795.

Manders, K., van de Poll-Franse, L.V., Creemers, G-J., Vreugdenhil, G., van der Sangen, MJC, Nieuwenhuijzen, GAP, et al., 2006. Clinical management of women with metastatic breast cancer: A descriptive study according to age group. BMC Cancer 6, 179. https://doi.org/10.1186/1471-2407-6-179.

Morgan-Parkes, J.H., 1995. Metastases: Mechanisms, pathways, and cascades. Am. J. Roentgenol. 164, 1075–1082. https://doi.org/10.2214/ajr.164.5.7717206.

Moshe Bocher, Ira M. Blevis; Leonid Tsukerman; Yigal Shrem; Gil Kovalski; Lana Volokh. A fast cardiac gamma camera with dynamic SPECT capabilities: Design, system validation and future potential. 2010;37:1887–1902. https://doi:10.1007/s00259-010-1488-z.

Palumbo, B., Sivolella, S., Palumbo, I., Liberati, A.M., Palumbo, R., 2005. 67Ga-SPECT/CT with a hybrid system in the clinical management of lymphoma. Eur. J. Nucl. Med. Mol. Imaging 32, 1011–1017. https://doi.org/10.1007/s00259-005-1788-x.

Patel, C.N., Salahudeen, H.M., Lansdown, M., Scarsbrook, A.F., 2010. Clinical utility of ultrasound and 99mTc sestamibi SPECT/CT for preoperative localization of parathyroid adenoma in patients with primary hyperparathyroidism. Clin. Radiol. 65, 278–287.

Prekeges, J., 2011. Nuclear Medicine Instrumentation, 2nd edition. Jones and Bartlett Publishers.

Rager, O., Nkoulou, R., Exquis, N., Garibotto, V., Tabouret-Viaud, C., Zaidi, H., Amzalag, G., Lee-Felker, S.A., Zilli, T., Ratib, O., 2017. Whole-body SPECT/CT versus planar bone scan with targeted SPECT/CT for metastatic workup. Biomed. Res. Int. 2017, 7039406.

Rajasekaran, A.K., Anilkumar, G., Christiansen, J.J., 2005. Is prostate-specific membrane antigen a multifunctional protein? Am. J. Physiol. Cell Physiol. 288, C975–C981. https://doi.org/10.1152/ajpcell.00506.2004.

Roach, P.J., Schembri, G.P., Ho Shon, I.A., Bailey, E.A., Bailey, D.L., 2006. SPECT/CT imaging using a spiral CT scanner for anatomical localization: Impact on diagnostic accuracy and reporter confidence in clinical practice. Nucl. Med. Commun. 27, 977–987. https://doi:10.1097/01.mnm.0000243372.26507.e7 PMID: 17088684.

Sengupta, S., Asha Krishnan, M., Chattopadhyay, S., Chelvam, V., 2019. Comparison of prostate specific membrane antigen ligands in clinical translation research for diagnosis of prostate cancer. Cancer Rep. 2, 1–22. https://doi.org/10.1002/cnr2.1169.

Seo, Y., Mari, C., Hasegawa, B.H., 2008. Technological development and advances in single-photon emission computed tomography/computed tomography. Semin. Nucl. Med. 38, 177–198. https://doi.org/10.1053/j.semnuclmed.2008.01.001.

Shenoy, P., Maggioncalda, A., Malik, N., Flowers, C.R., 2011. Incidence patterns and outcomes for Hodgkin lymphoma patients in the United States. Adv. Hematol. 2011, 725219. https://doi.org/10.1155/2011/725219.

Siegel, R.L., Miller, K.D., Jemal, A., 2019. Cancer statistics, 2019. CA Cancer J. Clin. 69, 7–34. https://doi.org/10.3322/caac.21551.

Tricoli, J.V, Schoenfeldt, M., Conley, B.A., Tricoli, J.V., Schoenfeldt, M., Conley, B.A., 2004. Detection of prostate cancer and predicting progression : Current and future diagnostic. Clin. Cancer Res. 10, 3943–3953. https://doi.org/10.1158/1078-0432.CCR-03-0200.

Uccella, S., Morris, J.M., Bakkum-Gamez, J.N., Keeney, G.L., Podratz, K.C., Mariani, A., 2013. Bone metastases in endometrial cancer: Report on 19 patients and review of the medical literature. Gynecol. Oncol. 130, 474–482. https://doi.org/10.1016/j.ygyno.2013.05.010.

Zhang, C., Liu, L., Tao, F., Guo, X., Feng, G., Chen, F., et al., 2018. Bone metastases pattern in newly diagnosed metastatic bladder cancer: A population-based study. J. Cancer 9, 4706–4711. https://doi.org/10.7150/jca.28706.

Mammography— sentinel of breast cancer management

Joseph Thomas[a], Amulya Cherukumudi[b], Pramod V[c]

[a]Department of Plastic Surgery, Kasturba Medical College Manipal, Manipal Academy of Higher Education, Manipal, Karnataka
[b]Consultant Surgeon, The Bangalore Hospital, Bangalore, Karnataka
[c]Consultant Radiologist, Delta Diagnostic Centre, Bangalore, Karnataka

7.1 Introduction

Breast cancer is the most common form of cancer diagnosed in women today: accounting for one in every four cancers (Bray et al., 2018). Over the last century, its management has evolved, significantly improving outcomes. Mammography has been pivotal to the evolution in detection and treatment of breast cancer being extensively used for both screening as well as confirmation of diagnosis. It is a simple radiographic technique using low-energy x-rays to examine human breasts for the presence of disease, particularly carcinoma.

Mammography is classified based on its role in the management of breast disorders:

1. Screening mammography (SM)
2. Diagnostic mammography (DM)

SM is the most frequently used modality for detecting early breast cancer in asymptomatic women. Due to its unique anatomy and external location, diseases of the breast, particularly tumours, are amenable to early detection. Mammography takes advantage of this to provide a safe, effective, and economic option for breast cancer screening. It is also the only imaging modality proven to significantly lower mortality from breast cancer (Marmot et al., 2013).

As per the American Cancer Society screening guidelines, women above 45 years of age benefit from a yearly mammogram (Oeffinger et al., 2015); those between 40 and 44 years can get them done annually at their discretion. For those women who have genetic predisposition, clinical breast examination combined with mammography is advised even earlier (Oeffinger et al., 2015).

Diagnostic mammography, on the other hand, is performed to confirm the diagnosis of breast cancer. It is typically a part of "the triple test", which includes clinical evaluation, pathologic evaluation, and radiologic evaluation using mammography or ultrasonography.

Biomedical Imaging Instrumentation. DOI: https://doi.org/10.1016/B978-0-323-85650-8.00002-4

Despite the arrival of several newer modalities of imaging over the years, it has stayed relevant and even adopted several of them as adjuvants, greatly enhancing its value in the management of breast cancer.

This chapter deals with mammography's evolution as an irreplaceable tool for breast cancer management, the science behind it, its adoption in clinical practice, the many adjuvants, the future trends, and its rival alternatives.

7.2 Evolution of mammography

In 1913, German surgeon Albert Salomon published in a monogram a series of mammography studies performed on 3000 mastectomy patients, comparing x-rays of the breasts to the removed tissue. This study demonstrated the possibility of correlating radiologic, macroscopic, and microscopic anatomy of breast diseases with differentiation between benign and malignant entities (Gold, 1992; Sharyl & Henderson, 2001).

"A Roentgenologic Study of the Breast," published in 1930 by American physician and radiologist Stafford L. Warren, detailed a study on stereoscopic x-ray images of breast tissue during pregnancy and mastitis (Sara, 2021). Warren was able to predict cancer's presence with very high accuracy among the women who subsequently went in for surgery.

Raul Leborgne, a Uruguayan radiologist, revitalized the interest in mammography in the 1950s, calling for attention to the need for technical qualification of patients positioning and radiologic parameters to be adopted. He was a pioneer in the enhancement of imaging quality, putting particular emphasis on the differential diagnosis between benign and malignant calcifications and elaborating on the need for breast compression during imaging (Leborgne, 1951).

In the late 1950s, Robert Egan at the M.D. Anderson Cancer Center combined the use of low-voltage technique with high milliamperage, and unique single emulsion films developed by Kodak to devise a method of screening mammography for the first time. The Egan technique improved the ability of physicians to detect calcification in breast tissue (Medichand Martel, 2006; Rebecca, 2016).

Mammography as a screening technique was popularized further, following a 1966 study led by Philip Strax demonstrating the impact of mammograms on mortality and treatment (Barron, 2003; Shapiro et al., 1966).

The American College of Radiology (ACR) established committees and centers for training at a countrywide level, and soon formed the ACR Mammography Committee (Feig, 1987; Gold, 1992; Kimme-Smith, 1992).

Before long, Physicains surgeons and radiologists adopted it as the standard of care in breast cancer imaging for both screening as well as diagnosis. Over time, scientists incorporated modifications and adjuvant techniques and technologies such as digital imaging. These further enhance the efficacy and their utility, cementing mammography to the management of breast cancer management for the near foreseeable future.

7.3 Dissecting the mammography machine

A. **Conventional mammography** (Faridah, 2008; Kuzmiak et al., 2005). Mammography, otherwise referred to as "early breast radiography," is essentially an x-ray of compressed breast tissue taken from multiple orientations. Architectural irregularities and calcific changes are looked for and analyzed by the radiologist or physician.

The various components of the machine with a comparative assessment of conventional radiographic devices are described here.

i. **Anode tube.** Early mammographic machines used tungsten similar to conventional x-ray machines. However, manufacturers make present-day machines of molybdenum or dual material, created by embedding rhodium to give a characteristic spectrum optimized for breast imaging.

ii. **Filters.** They function to filter out x-ray beams, in order to reduce the extent of exposure to patients. Often, molybdenum is preferred for mammography, in contrast to the aluminium or aluminium equivalent filters in conventional radiography machines. Some modern devices allow the operator to toggle between molybdenum and rhodium filters to vary the radiation spectrum for specific breast conditions. These filters have an added advantage over those used in conventional radiography by enhancing the mammography's contrast sensitivity.

iii. **Focal spots.** It is the area of the anode surface which receives the beam of electrons from the cathode; the apparent source of x-rays. Most typical mammography machines use two focal spots, which can be selected depending on the suspected breast lesion. In the past, large focal spots were used (by Egan) to optimize the mammography's primitive design. However, this generated a geometric disparity, which was compensated by using a greater focal spot-to-film distance. The x-ray tube's focal spots for a mammography are smaller than conventional radiographs to reduce the blurring and improve visualization and identification of small calcifications. Of these two focal spots, the smaller one is required to magnify the images being generated.

iv. **Breast compression unit.** This singular step revolutionized x-ray for breast imaging by greatly enhancing its sensitivity. While this improves the efficacy, it is also the most common deterrent for women to undergo this procedure, as it may cause pain and discomfort (Fig. 7.1). Two views are obtained in a mammogram based on the type of compression (i.e., mediolateral, craniocaudal). The advantages with compression of the breast tissue are as follows:
1. Uniformity in the breast thickness
2. Restriction of breathing movements of the patient, which minimizes blurring of the image rendered

FIG. 7.1

Mammography machine with compression plates visualized.

    3. Reduction in scattering of radiation and improvement in contrast sensitivity

    4. A lower dose of radiation required (from 120 kVp to 20 kVp)

v. Grid. This helps in absorbing the excess scattered x-ray beams while improving contrast sensitivity. The grid is contained in a Bucky device that moves it during the x-ray exposure to blur and reduce the grid lines' visibility.

vi. Receptor. This can be a film or a screen, which picks up the x-ray beams and produces the images that form the mammogram.

B. Digital mammography. The first full-field digital receptor was launched in 2005 after US Food and Drug Administration approval and has subsequently been implemented in hospitals across the world (Bahl et al., 2019). It has replaced conventional mammography as the gold standard for breast screening and evaluation. Digitization has resulted in significant contrast transfer from breast to film, maximizing the contrast sensitivity (Zanca et al. 2009). Three crucial features of digital mammography are as follows:

    i. Digital receptor dynamic range. Digitization ensures a wide range of possible exposures. This allows the receptor to tolerate variation in direction and a greater fidelity in the contrast sensitivity (Bahl et al., 2019).

    ii. Digital image processing. The image obtained from the machine undergoes processing, which improves quality and visibility. Among these processing techniques, contrast processing is the most common. It allows the digital mammogram images to appear more like their conventional counterparts (Bahl et al., 2019). There are several algorithms available for the processing of images to improve the clarity of the image and highlight the areas of interest. For instance, if there is a suspicious area of microcalcification, image processing can enhance that particular portion (Zanca et al., 2012). Some well-known manufacturers in the field of image processing algorithms are OpView, Raffaello, and Sigmoid.

iii. **Windowing.** It is the process of selecting one segment of the total pixel value range (the wide dynamic range of the receptors) and then displaying the pixel values within that segment over the full brightness (shades of gray) ranging from white to black. This technique optimizes the contrast and visibility of the surrounding structures in a given digital image (Bahl et al., 2019).

7.4 **How a mammogram is captured**

The **cathode-anode axis** is oriented perpendicular to the chest wall, with the cathode on the chest wall side and the anode on the side of the nipple. This anode heel effect reduces tube output on the anode side and decreases the equipment bulk near the patient's head for a more comfortable positioning (Bluekens et al., 2012; Zanca et al., 2012). The target focus is placed directly above the chest wall edge to minimize beam divergence into the patient (Bluekens et al., 2012). To achieve the required spatial resolution, the mammography tube must have a small focal spot. The smaller focal spot tube current is maintained at 40 to 100 mA and voltage potential is 26 to 35 kV (Berkhout et al., 2004). The focal spot(s) is usually placed above while the receptor is kept below.

The breast is compressed between two surfaces. Typically, one is stationary while the other moves to compress and spread the breast out, to aid in eliminating motion and blurring of the image. The compression may be uncomfortable but should not be painful and must be applied for only a few seconds. The image is taken, and compression is released.

Two views are frequently used: **craniocaudal** and **mediolateral**. The direction of compression, too, is changed accordingly. Images are then sent for processing and evaluation.

7.5 **Deciphering a mammogram**

Once the image is processed, radiologists and clinicians assess the images for any features suggestive of benign or malignant lesions (Fig. 7.2). These findings are evaluated using **BI-RADS (Breast Imaging–Reporting and Data System)**, a risk assessment and quality assurance tool developed by the ACR that provides a widely accepted lexicon and reporting schema for imaging of the breast (Berkhout et al., 2004; Huda et al., 2016). This has dramatically improved mammography reporting quality, making it more objective and universally uniform (Berkhout et al., 2004). It has proven to be an efficient and cost-effective method to analyze and categorize mammograms (Huda et al., 2016).

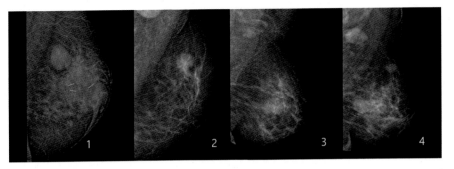

FIG. 7.2 Mammogram showing

1. Post-operative case with well defined radio-dense lesion, adjacent architectural distortion and enlarged axillary lymph nodes- suggestive of recurrence 2. A well-defined round radiodense lesion, no distortion or thickening, with normal appearing lymph nodes- BI-RADS 2 3. A well-defined radiodense lesion, with adjacent architectural distortion, clustered pleomorphic calcification extending to Retromammary space with fine lobulation, BIRADS 5 with multiple enlarged axillary lymph nodes (loss of lobulation and radiolucency) 4. A well-defined radiodense lesion, with adjacent architectural distortion, clustered pleomorphic calcification fine lobulation, BIRADS 5.

BI-RADS describes the breast composition or density, masses, calcifications, asymmetries, associated features, and location of the lesion (Magny et al., 2020).

1. Breast density is described as fatty, scattered, heterogeneously dense, and extremely dense.
2. If a mass is visualized, the shape, margin, and density need to be described in detail. The shape can be round, oval, or irregular. The margins can be circumscribed, obscured, microlobulated, indistinct, and spiculated. The mass density can be high density, equal density, low density, and fat containing. Irregular, high-density mass with spiculated margin increases the suspicion of malignancy.
3. Calcifications: Benign calcifications can be described as large rodlike, popcorn, coarse, vascular, and milk of calcium. Those that raise suspicion of malignancy are amorphous, fine-pleomorphic, and fine-linear branching.
4. Distribution of calcifications: This can be described as diffuse, regional, grouped, linear, and segmental.
5. There can be other features such as the presence of skin or nipple retraction, skin or trabecular thickening, and axillary adenopathy.
6. Other features observed on a mammogram are architectural distortion, intramammary lymph node, skin lesion, and solitary dilated duct.
7. Location: This is based on laterality, quadrant or clock face, depth, or distance from the nipple.

7.6 Limitations

Dense breasts and younger age groups are not ideal for mammography, since breast tissue density leads to artifacts and inaccurate reporting and a more significant number of false positives.

During the procedure, discomfort and pain lead to poorer patient compliance and participation in screening programs. Furthermore, overreporting through false positives may cause avoidable mental distress to many women (Sylvia et al., 2011).

Mammographs use ionizing radiation in the form of x-rays. Despite the low dosage, repeated tests could lead to radiation-induced side effects (National Research Council, 2006).

7.7 Recent advances in mammography

7.7.1 Contrast-enhanced mammography

Contrast-enhanced mammography (CEM) was developed to overcome the reduced sensitivity and specificity of digital mammograms in dense breast tissue (Cameron et al., 2012). Mammographic images are taken before and after injection of an iodine-based dye. Using a dual-energy technique that describes the difference in x-ray attenuation of breast tissue and contrast, it delineates areas of contrast agent uptake within the breast, which is often found in breast carcinoma (Cameron et al., 2012; Norman et al., 2020).

CEM images are comparable to breast magnetic resonance imaging (MRI) in breast imaging. The anatomic and physiologic details of CEM have also shown superior sensitivity and specificity for breast carcinoma diagnosis with digital mammography alone (Cameron et al., 2012; Perry et al., 2018).

Recombined images are formed by the subtraction of low-energy images from high-energy images, which cancels signal from the background of breast tissue and highlights areas of contrast uptake (Perry et al., 2018).

7.7.2 Three-dimensional mammography

Three-dimensional (3D) mammography, also known as digital breast tomosynthesis (DBT), tomosynthesis, and 3D breast imaging, creates a 3D image. When used in addition to usual mammography it results in greater overall sensitivity (Hodgson et al., 2016).

7.7.3 Photon counting mammography

Photon counting mammography reduces the dose of x-ray presented to the patient by approximately 40%, while maintaining image quality at an equal or higher level (Weigel et al., 2014). It has subsequently developed to enable spectral imaging (Berglund et al., 2014; Sudhir et al., 2020) with the possibility to further improve image quality (Berglund et al., 2014), to distinguish between different tissue types (Fredenberg et al., 2018), and to measure breast density (Johansson et al., 2017).

7.7.4 Galactography

Galactography, also known as breast ductography, is used to visualize the milk ducts by injecting a radiopaque substance into the duct system. This technique is indicated in the presence of suspicious nipple discharge.

7.8 Alternatives

Alternative modalities, some established and several under research, are mentioned in this section. We may see some of them individually or as hybrid systems, replace mammography in the screening and diagnosis of breast cancer (Sheth & Giger, 2019; Sivaramakrishna, 2005).

- Ultrasonography
- Magnetic resonance mammogram
- Positron emission mammogram
- Infrared-based thermography
- Hybrid techniques such as full field digital mammography and ultrasound (FFDMUS)
- Electrical impedance scanning (EIS)
- Microwave breast imaging

7.9 Conclusion

Mammography, conventional or otherwise, has revolutionized breast cancer screening and diagnosis. With improvements in the machine and the image rendering, early diagnosis and treatment can be achieved with minimal radiation exposure while remaining cost-effective. It has managed to incorporate advances in radiology and imaging technology to stay relevant despite the emergence of several alternatives. For the near foreseeable future, mammography will remain integral to the screening and diagnosis of breast cancer across the globe.

References

Alghaib, H.A., Scott, M., Adhami, R.R., 2016. An overview of mammogram analysis. IEEE Potentials 35 (6), 21–28.

Bahl, M., Pinnamaneni, N., Mercaldo, S., McCarthy, A.M., Lehman, C.D., 2019. Digital 2D versus tomosynthesis screening mammography among women aged 65 and older in the United States. Radiology 291 (3), 181637.

Berglund, J., Johansson, H., Lundqvist, M., Cederström, B., Fredenberg, E., 2014. Energy weighting improves dose efficiency in clinical practice: Implementation on a spectral photon-counting mammography system. Journal of Medical Imaging 1 (3), 031003.

Berkhout, E., Beuger, D.A., Sanderink, G., van der Stelt, P.F., 2004. The dynamic range of digital radiographic systems: Dose reduction or risk of overexposure? Dentomaxillofacial Radiology 33 (1), 1–5.

Bluekens, A.M.J., Holland, R., Karssemeijer, N., Broeders, M.J.M., den Heeten, G.J., 2012. Comparison of digital screening mammography and screen-film mammography in the early detection of clinically relevant cancers: A multicenter study. Radiology 265 (3), 707–714.

Bray, F., Ferlay, J., Soerjomataram, I., Siegel, R.L., Torre, L.A., Jemal, A., 2018. Global cancer statistics 2018: GLOBOCAN estimates of incidence and mortality worldwide for 36 cancers in 185 countries. A Cancer Journal for Clinicians 68 (6), 394–424.

Cameron, D.A., Dewar, J.A., Thompson, S.G., Wilcox, M., 2012. The benefits and harms of breast cancer screening: An independent review. Lancet 380 (9855), 1778–1786.

Faridah, Y., 2008. Digital versus screen film mammography: A clinical comparison. Biomedical Imaging and Intervention Journal 4 (4), e31.

Feig, S.A., 1987. Mammography equipment: principles, features, selection. Radiologic Clinics of North America 25 (5), 897–911.

Fredenberg, E., Willsher, P., Moa, E., Dance, D.R., Young, K.C., Wallis, M.G., 2018. Measurement of breast-tissue x-ray attenuation by spectral imaging: Fresh and fixed normal and malignant tissue. Physics in Medicine & Biology 63 (23), 5003.

Gold, R.H., 1992. The evolution of mammography. Radiologic Clinics of North America 30 (1), 1–19.

Heywang-Köbrunner, S.H., Hacker, A., Sedlacek, S., 2011. Advantages and disadvantages of mammography screening. Breast Care (Basel) 6 (3), 199–207.

Hodgson, R., Heywang-Köbrunner, S.H., Harvey, S.C., Edwards, M., Shaikh, J., Arber, M., Glanville, J., 2016. Systematic review of 3D mammography for breast cancer screening. Breast 27, 52–61.

Hubbard R.A., O'Meara E.S., Henderson L.M., et al., 2016. Multilevel factors associated with long-term adherence to screening mammography in older women in the U.S. Preventive Medicine 89, 169-177. doi:10.1016/j.ypmed.2016.05.034.

Huda, S., Ali, T., Nanji, K., Cassum, S., 2016. Perceptions of Undergraduate Nursing Students Regarding Active Learning Strategies, and Benefits of Active Learning. International Journal of Nursing Education. 8, 193. doi: 10.5958/0974-9357.2016.00151.3.

Institute of Medicine (US) and National Research Council (US) Committee on the Early Detection of Breast Cancer, 2001. Mammography and beyond: Developing technologies for the early detection of breast cancer: A non-technical summary. National Academies Press.

Johansson, H., von Tiedemann, M., Erhard, K., Heese, H., Ding, H., Molloi, S., Fredenberg, E., 2017. Breast-density measurement using photon-counting spectral mammography. Medical Physics 44 (7), 3579–3593.

Kimme-Smith, C., 1992. New and future developments in screen-film mammography equipment and techniques. Radiologic Clinics of North America 30 (1), 55–66.

Kuzmiak, C.M., Pisano, E.D., Cole, E.B., Zeng, D., Burns, C.B., Roberto, C., Pavic, D., Lee, Y., Seo, B.K., Koomen, M., Washburn, D., 2005. Comparison of full-field digital mammography to screen-film mammography with respect to contrast and spatial resolution in tissue equivalent breast phantoms. Medical Physics 32 (10), 3144–3150.

Leborgne, R., 1951. Diagnosis of tumors of the breast by simple roentgenography; calcifications in carcinomas. The American Journal of Roentgenology and Radium Therapy 65 (1), 1–11.

Lerner, B.H., 2003. To see today with the eyes of tomorrow: A history of screening mammography. Canadian Bulletin of Medical History 20 (2), 299–321. doi:10.3138/cbmh.20.2.299.

Magny, S.J., Shikhman, R., Keppke, A.L., 2020. Breast imaging reporting and data system. StatPearls Publishing.

Marmot M.G., Altman D.G., Cameron D.A., Dewar J.A., Thompson S.G., Wilcox M., 2013. The benefits and harms of breast cancer screening: An independent review. British Journal of Cancer 108(11), 2205-2240. doi:10.1038/bjc.2013.177.

Medich D.C., Martel C., 2006. Medical Health Physics. Health Physics Society. Summer School. Medical Physics Publishing.

National Research Council, 2006. Health risks from exposure to low levels of ionizing radiation: BEIR VII phase 2. The National Academies Press.

Norman, C., Lafaurie, G., Uhercik, M., Kasem, A., Sinha, P., 2021. Novel wire-free techniques for localization of impalpable breast lesions—a review of current options. Breast Journal 27 (2), 141–148. doi:10.1111/tbj.14146.

Oeffinger, K.C., Fontham, E.T., Etzioni, R., Herzig, A., Michaelson, J.S., Shih, Y.C., Walter, L.C., Church, T.R., Flowers, C.R., LaMonte, S.J., Wolf, A.M., DeSantis, C., Lortet-Tieulent, J., Andrews, K., Manassaram-Baptiste, D., Saslow, D., Smith, R.A., Brawley, O. W., Wender, R., & American Cancer Society, 2015. Breast Cancer Screening for Women at Average Risk: 2015 Guideline Update From the American Cancer Society. Journal of American Medical Association, 314 (15), 1599–1614. https://doi.org/10.1001/jama.2015.12783.

Perry, H., Phillips, J., Dialani, V., Slanetz, P.J., Fein-Zachary, V.J., Karimova, E.J., Mehta, T.S., 2018. Contrast-enhanced mammography: A systematic guide to interpretation and reporting. American Journal of Roentgenology 212 (1), 222–231.

Screening, Independent UK Panel on Breast Cancer, 2012. The benefits and harms of breast cancer screening: An independent review. Lancet 380 (9855), 1778–1786.

Shapiro, S., Strax, P., Venet, L., 1966. Evaluation of periodic breast cancer screening with mammography. Journal of American Medical Association 195 (9), 731–738.

Sharyl, J., Nass, S.J., Henderson, I.C., et al., eds. Mammography and beyond: Developing technologies for the early detection of breast cancer. National Research Council; Institute of Medicine; Division on Earth and Life Studies; National Cancer Policy Board; Committee on Technologies for the Early Detection of Breast Cancer, 2001.

Sheth, D., Giger, M.L., 2019. Artificial intelligence in the interpretation of breast cancer on MRI. Journal of Magnetic Resonance Imaging. doi:10.1002/jmri.26878.

Sivaramakrishna, R., 2005. Imaging techniques alternative to mammography for early detection of breast cancer. Technology in Cancer Research & Treatment 4 (1), 3.

Sudhir, R., Sannapareddy, K., Potlapalli, A., Krishnamurthy, P.B., 2020. Diagnostic accuracy of contrast-enhanced digital mammography in breast cancer detection in comparison to tomosynthesis, synthetic 2D mammography and tomosynthesis combined with ultrasound in women with dense breast. The British Journal of Radiology, 20201046. doi:10.1259/bjr.20201046.

Turbow S.D., White M.C., Breslau E.S., Sabatino S.A., 2021. Mammography use and breast cancer incidence among older U.S. women. Breast Cancer Research and Treatment 188(1), 307-316. doi:10.1007/s10549-021-06160-4.

Warren, S.L., 1930. A roentgenologic study of the breast. The American Journal of Roentgenology and Radium Therapy 24, 113–124.

Weigel, S., Berkemeyer, S., Girnus, R., Sommer, A., Lenzen, H., Heindel, W., 2014. Digital mammography screening with photon-counting technique: Can a high diagnostic performance be realized at low mean glandular dose? Radiology 271 (2), 345–355.

Zanca, F., Jacobs, J., Van Ongeval, C., Claus, F., Celis, V., Geniets, C., Provost, V., Pauwels, H., Marchal, G., Bosmans, H., 2009. Evaluation of clinical image processing algorithms used in digital mammography. Medical Physics 36 (3), 765–775.

Zanca, F., Van Ongeval, C., Claus, F., Jacobs, J., Oyen, R., Bosmans, H., 2012. Comparison of visual grading and free-response ROC analyses for assessment of image-processing algorithms in digital mammography. The British Journal of Radiology 85 (1020), e1233–e1241.

Hyperspectral imaging: Current and potential clinical applications

Sakir Ahmed[a], Prajna Anirvan[b], Priyanku Pratik Sharma[c], Manmath Kumar Das[d]

[a]*Department of Clinical Immunology & Rheumatology, Kalinga Institute of Medical Sciences, KIIT University, Bhubaneswar*
[b]*Department of Gastroenterology, S.C.B Medical College, Cuttack*
[c]*Department of Urology, Dispur Hospital, Guwahati*
[d]*Kalinga Institute of Medical Sciences, KIIT University, Bhubaneswar*

8.1 Introduction

Hyperspectral imaging (HSI) can be considered a form of spectroscopy with data being collected across numerous narrow spectral bands. It has the advantage of collating a large volume of imaging data that can be captured via conventional cameras or seen by human eyes. Thus it can provide information on constituent elements, energy states, and a multitude of other aspects.

HSI first appeared in remote sensing satellites and has evolved since. From astronomy, geophysics, and defence surveillance, it has spread to agriculture, weather forecasting, and then biomedical imaging and molecular biology. In the medical field, potential applications are for noninvasive diagnosis and guiding surgical procedures (Lu & Fei, 2014). Currently, most of the applications are potential, and only a few have translated to clinical use. However, things are likely to change very rapidly as the technologies to image and to process the big data involved become cheaper and more accessible. Many more applications will likely be added to this list once HSI is available in mainstream patient management.

8.2 Concept

A classical camera that captures colored images does it by filtering visible light into three primary colors (red, blue, and green). Each is captured alternatingly at high frequency. Higher quality cameras use a beam splitter and three sensors each with a different color filter. Thus the entire spectral information is binned into three variables. This is a pragmatic and effective method to store and process a large amount of (visual) information. The quality of data is minimally affected while cutting down raw data into manageable sizes.

Biomedical Imaging Instrumentation. DOI: https://doi.org/10.1016/B978-0-323-85650-8.00003-6

An example of this data condensation can be understood by users of the digital single-lens reflex (DSLR) camera. Whereas the final (compressed) picture in JPG format occupies much less space, the complete data initially stored by the sensor in RAW format would be approximately 10 times larger than the corresponding JPG. All cameras automatically compress large amounts of visual data into manageable bins. In most cases, this is sufficient. But an amateur photographer who has attempted to take a picture of a dancing peacock would realize the picture looks very dull as compared to the real bird because the vibrant colors of the bird have been binned into a few significant components. One way around (used by professional photographers) is to tweak the binning of the RAW files. But the ideal way would be to have multiple sensors to capture the entire visible spectrum instead of just the RBG sensors.

8.3 **Principles**

Basic spectral imaging uses cameras that split visible light into three board bands: long wavelengths as red, medium wavelengths as green, and short wavelengths as blue. Multispectral imaging bins visual data into 5 to 20 wide bands. By convention, HSI bins into more than 20 equally distributed narrow bands of wavelengths (Fig. 8.1). Hyperspectral imaging utilizes optical spectroscopy with two-dimensional (2D) object

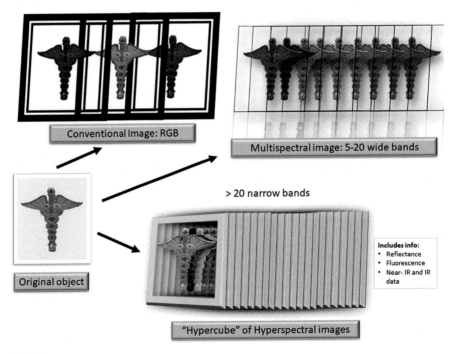

FIG. 8.1

Differences between conventional RGB (red-green-blue) image, multispectral image, and hyperspectral image.

mapping. Thus the outcome is 3D hypercube with two dimensions recording the spatial location (location of the pixel) and the third dimension recording the spectral information, including a range of wavelengths (Pierce et al., 2012). Data acquisition through HSI provides detailed analysis of spatial spectrum variation, thereby providing greater diagnostic detail. This resolution may not be limited to the visible spectrum but also include infrared and ultraviolet spectra. Also, the third dimension can record multiple pixel parameters such as reflectance, fluorescence, and transmission.

8.4 Techniques

All HSI image capturing techniques can be broadly classified under four classes:

A. Spatial scanning: This involves focusing on a small area and obtaining a complete slit spectrum for that area. This is done by projecting a strip (or a line of the image) and dispersing its spectra with the help of prisms. Thus the complete image is screened line by line. This is done by a push-broom sensor (Fig. 8.2). If instead of scanning per line, pointwise scanning is done using point aperture instead of a slit, called a whisk-broom scanner.

B. Spectral scanning: Here the image is captured by multiple monochromic sensors. The scene is scanned by numerous filters (either fixed or tunable) consecutively.

C. Nonscanning: Also called snapshot HSI, a single device captures the entire data cube at one time. These systems acquire mostly chemical information and not true spectral information. Thus they are not amenable to postprocessing or reanalysis.

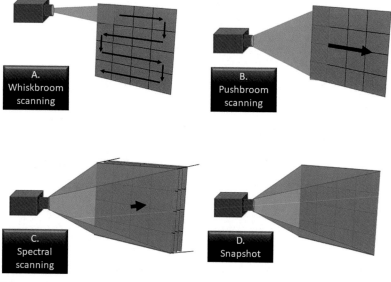

FIG. 8.2

Types of image acquisition for hyperspectral imaging.

D. Spatiospectral scanning: This is a combined attempt to overcome the individual disadvantages of spatial and spectral scanning. Each sensor captures a wavelength-encoded spatial map, thus a 3D map. Moving the camera or the slit allows for scanning, including another dimension to the hypercube.

Newer cost-effective approaches may include using regular digital cameras with diffraction filters and machine learning to construct the HSI image (Fauvel et al., 2013; Toivonen et al., 2021).

8.5 Analysis

Initial analysis was based on approaches such as principle component analysis and least squares–based mixed analysis (Heinz & Chang, 2001). Currently strategies are greatly influenced by advances in big data analysis such as use of machine learning and neural networks. The details are beyond the scope of this chapter, and the interested reader is referred to comprehensive reviews on the topic (Plaza et al., 2009).

8.6 Biomedical uses

Biomedical uses can be summarized in the following table.

No	Principle	Example
1.	Delineating vital versus nonvital tissue	In surgeries such as tumor resection or vital organ (kidney/liver) resection
2.	Imaging/quantification of inflamed tissue	For diagnoses such as arthritis or low blood flow in diseases of blood vessels
3.	Visualizing abnormal tissue	Role in cancer screening: increases yield of screening cytology or biopsies
4.	3D imaging of in vitro models	Role in increasing resolution of confocal microscopy
5.	Use of specific spectro-scopic data	For rapid clinical diagnoses such as hemoglobin estimation, tissue carbon dioxide retention or perfusion

These will be explored in the context of the following subsections:

1. External scanning (skin, joints, blood vessels)
2. Endoscopic applications
3. Applications in surgery
4. Applications for imaging the eye
5. In vitro use

8.7 Use externally to the human body

External use can be to scan for systemic parameters, for diagnosis of dermatologic disorders, inflammatory arthritis, or vascular flow.

8.7.1 Monitoring of patients in critical care

For patients in the emergency ward or critical care units, tissue perfusion should be monitored. Shock due to various causes leads to inadequate blood circulation and hence decline in tissue perfusion. Thus quantification of tissue perfusion is the best monitor for deterioration or improvement in patients in shock. HIS has the potential to inform the clinician about blood oxygen saturation, total hemoglobin distribution, and tissue water content, all of which are relevant for immediate decision making (Sicher et al., 2018).

8.7.2 Diagnosis of skin cancers

HSI can be used to virtually biopsy the skin into five approximate layers accessing tissue density and perfusion (Marotz et al., 2019). HSI-acquired data can be processed by convoluted neural networks to automatically differentiate skin cancers such as melanoma from benign skin conditions (Hirano et al., 2020). Such systems have also been developed for squamous cell carcinomas of the skin (Liu et al., 2020). One such system using 125 spectral band HSI has shown sensitivity and specificity results of 87.5% and 100%, respectively (Leon et al., 2020).

8.7.3 Diagnosis and monitoring of other skin lesions

Ulcers can occur due to various causes such as trauma, venous stasis, low blood flow, loss of protective pain responses (as in diabetes), or rarely cancer. HSI can help differentiate the underlying cause, map the extent of damage, and help in monitoring response to therapy (Zimmermann & Curtis, 2019). It can also help in planning debridement of ulcers and other wounds by delineating dead, dying, and vital tissue (Daeschlein et al., 2017).

Diabetic foot ulceration is a dreaded complication of this disease, having a high risk of posing serious consequences like affected limb amputation. The microcirculatory pathologies, in addition to extensive vessel disease, are responsible for these complications. Different studies have revealed that HSI has the ability to predict diabetic foot ulcer with high sensitivity and specificity. The studies have used methods to produce a map of oxyhemoglobin and deoxyhemoglobin concentrations in the dermis. They can predict a developing pathology by mapping the ulcerative mechanisms.

Fungal infections of the skin are prevalent and are commonly caused by three genera called the dermatophytes. his-based applications can aid in spot diagnosis without the need for an expert dermatologist or microbiologist (Ye et al., 2018).

8.7.4 Screening for arthritis

Inflammatory arthritis, if untreated, can lead to rapid joint destruction and lifelong deformities. The limited number of rheumatologists worldwide make the diagnosis of such arthritis difficult. HSI systems can be developed that can detect the presence of inflammatory arthritis (Milanic et al., 2015). This has tremendous potential not only for screening of arthritis but also for monitoring and treating to target.

8.7.5 Screening for vascular diseases

Diseases of the vessels can be due to atherosclerosis or deposition of lipid in the vessel wall that induces hypertrophy of the muscle coat of the vessel. This leads to luminal compromise. Atherosclerosis is ubiquitous in all humans and increases with age and presence of various comorbidities. Then there are other diseases of vessels such as excess vasospasm in presence of cold (known as Raynaud phenomenon) or an uncommon group of diseases called vasculitis. All these can lead to regional loss of perfusion that can be picked up and quantified by HIS (Allen & Howell, 2014).

8.8 Use in endoscopies

8.8.1 Upper and lower gastrointestinal endoscopy

Endoscopic procedures to image both the upper and lower digestive tracts are frequently used for cancer screening and surveillance programs. Conventional endoscopes use white light—the so-called white light endoscopy (WLE)—where the reflected image is captured using a normal camera, which detects only the red, green, and blue colors, similar to the human eye. To overcome these barriers, chromoendoscopy and narrow band imaging (NBI) have evolved, which make use of chemical staining and mucosal appearance under certain specific wavelengths of light. NBI is a type of multispectral imaging that utilizes light of two distinct wavelengths, blue (415 nm) and green (540 nm). However, these techniques are not foolproof and are liable to errors more so in the detection of precancerous lesions of the gastrointestinal tract. The application of HSI to the field of medical science, and especially gastroenterology, is envisaged to improve cancer detection and lesion characterization.

There are, however, limitations to the application of HSI to the field of endoscopy in current gastroenterologic practice. The complex instrumentation required in HSI image acquisition and the huge geometric variations commonly seen during routine endoscopy are hindrances to the incorporation of HSI to endoscopy. In addition, the use of HSI is limited to the ex vivo study of tissue. In vivo esophageal application is still in its infantile stage owing to technical difficulties in the design of prevalent endoscopes.

There have been efforts to design endoscopes that could satisfy the requirements of HSI and result in clinical translation (Kiyotoki et al., 2013). HSI models have used tunable filters to convert WLE images to HSI images (Yoon et al., 2019). Endoscopically resected gastric and duodenal neoplasms have been evaluated through his, and differences in spectral reflectance between normal and malignant mucosa have been found ex vivo (Kiyotoki et al., 2013). Prior to the application of HSI, multispectral imaging has been applied to the detection of gastric carcinomas using upper GI endoscopy by modification of the standard Olympus endoscopy system (Hohmann et al., 2017). By using line-scanning HSI and reconstructing a hypercube, a method for data acquisition has been devised and has been technically validated in simulations. This hyperspectral endoscope model can be inserted through the accessory channel of a conventional endoscope, thus making clinical application of such technology possible in the near future (Yoon et al., 2019). Identification of precancerous lesions and intraepithelial papillary capillary loops, which are changes in surface mucosal blood vessels, can be done with HSI; this might, in future, be extrapolated to capsule endoscopy, too (Wu et al., 2018). Hyperspectral systems have also been evaluated in colonoscopy to aid in the detection of colorectal cancers (Kumashiro et al., 2016). Comparing between normal mucosa and neoplastic mucosa, this model found an increase in spectral absorption rates at 525 nm. Discrimination between normal and neoplastic tissues was reasonably good, with 72.5% sensitivity and 82.1% specificity, respectively.

8.8.2 Endoscopic retrograde chonlangiopancreatogram (ERCP) and choledochoscopy

As of yet, application of HSI to endoscopic visualization of the biliary tract through ERCP has not been tried. However, visualization of the biliary tree through HSI during laparoscopic surgery combining the near-infrared system with conventional laparoscope has been evaluated in experimental models (Zuzak et al., 2007). In porcine models, it has been possible to differentiate between gallbladder tissue and adjacent vascular tissue using HSI with a sensitivity and specificity of 98% (Livingston et al., 2009). Accurate delineation of the cystic duct in the hepatoduodenal ligament has also been achieved in such experimental models (Zuzak et al., 2007). The advantage of hyperspectral cholangiography lies in the fact that it obviates the need of any contrast agent. Although it would be premature to extrapolate these results to the human biliary system, validation studies are expected in the immediate future, thereby potentially leading to the application of HSI to choledochal imaging.

8.8.3 Video-assisted thoracic surgery (VATS)

The safety and feasibility of intraoperative near-infrared thoracoscopic localization using indocyanine green injection have been evaluated (Ujiie et al., 2017). Multispectral imaging has been evaluated in VATS. Use of fluorescence-assisted resection and exploration (FLARE) imaging system facilitates the identification of clear tumor margins and makes resection easier, taking care of nearby vital structures (DSouza et al., 2016). Real-time images and videos at 700 and 800 nm wavelengths are produced using FLARE (Gioux et al., 2010). Folate receptor–targeted intraoperative molecular imaging (FR-IMI) to reliably localize and resect pulmonary ground-glass opacities through minimally invasive thoracoscopic methods and VATS has shown promising results (Predina et al., 2018).

8.8.4 **Functional endoscopic sinus surgery and neuroendoscopy**

Functional endoscopic sinus surgery (FESS) and neuroendoscopy require fine surgical skills where visualization is of paramount importance. The ability of HSI to guide surgeons at multiple levels (tissue, cellular, and molecular) can be utilized in neuroendoscopy and FESS. Visualization during microsurgery and obtaining good images of tissue in the operative field where spillage of blood occurs are essential. HSI can help surgeons in this regard. Visualization of tissues under blood layers not accessible to the naked human eye has been tried through HSI (Monteiro et al., n.d.). HSI has the ability to help delineate the complex anatomy of tissues and vessels during surgery (Panasyuk et al., 2007). This could be used to assist head and neck surgeons during neuroendoscopy and FESS. In addition, in an in vivo validation database of hyperspectral images of brain tissue, it has been established that the differentiation of normal tissue from tumor tissue is possible through HSI (Fabelo et al., 2018). HSI also has the ability to perform real-time imaging, which would make it possible for neurosurgeons to confirm the presence of neoplasms and other pathologic lesions and plan surgery in the operation theater itself. The application of artificial intelligence (AI) to HSI is expected to open newer vistas in image analysis and establishing diagnosis (Halicek et al., 2017; Ma et al., 2017).

8.8.5 **Cystoscopy**

Real-time multispectral imaging (rMSI) has been used in the identification of bladder tumors. Combining white light imaging with rMSI, multiparametric cystoscopy has been performed successfully in bladder cancer (Kriegmair et al., 2020). Ex vivo studies on bladder cancer cells using multispectral imaging have demonstrated that cell spectra vary according to cancer stage with definite changes in the cell's composition (Jen et al., 2014). Although HSI models have not been directly applied in the cystoscopic detection of lesions, it is envisaged to play a major role shortly.

8.9 **Applications in surgery**

HSI has excellent potential for image-guided surgery and specifically for an intraoperative quantification of tissue perfusion. The relative reflectance of light within wavelengths 500 nm and 1000 nm allow differentiation between oxygenated and deoxygenated haemoglobin. It can play a major part in the following surgeries.

8.9.1 **Gastrointestinal anastomoses**

A very common situation faced by surgeons in their daily practice is the need for resection and anastomosis of bowel segments. The postanastomotic healing depends on a variety of factors, out of which the prime factor remains vascularity to the anastomotic site. Adequate vascularity means the water content in the tissue, the hemoglobin concentration, and the oxygen saturation will be maintained at optimum levels. Noninvasive HSI in this respect enables measurement of these crucial parameters intraoperatively and that, too, in a noncontact, radiation-free, and rapid manner. This proves to eliminate the risks of intraoperative administration of intravascular contrasts and radiation like the use of indocyanine dye with fluoroscopic assistance, which is used in certain scenarios (Jansen-Winkeln et al., 2018). Thus surgeries necessitating resection of benign and malignant tumors of the esophagus, stomach, pancreas, small and large bowel and subsequent anastomosis can benefit from the use of HSI technology in offering a relatively peaceful postoperative period to the concerned surgeon by predicting the ischemic conditioning effects and reduce postanastomotic complications.

8.9.2 Breast surgery

Tumor excision and reconstructive breast surgery are the major surgical procedures in this field. One of the primary goals of any cancer surgery is the achievement of wide excision with an R0 resection (i.e., absence of any histologically evident malignant tissue in the tumor bed or margins). The same dictum holds true for surgery of malignant breast tumors. HSI promises to have a higher sensitivity and specificity of detecting residual malignant tissue even in comparison to histopathologic analysis, thereby allowing the surgeon to have a brisk idea of his excisional efficacy, intraoperatively. HSI has the potential to identify tissues like tumors, blood vessels, muscle, and connective tissue. The algorithm based on spectral characteristics of tissue types helps to distinguish between tumor and normal cells and can identify residual tumour tissue up to 0.5 to 1 mm in size (Beletkaia et al., 2020).

In the field of breast surgery, HSI promises an extended role when it comes to reconstructive surgery. Breast cancer patients are treated not only with a focus on survival through excision (mastectomy) but also with a view to maintaining the quality of life through breast reconstructive surgery. Breast reconstruction with an autologous free deep inferior epigastric artery perforator (DIEP) flap being one of the preferred options after mastectomy, dynamic infrared thermography (DIRT) can be used as an efficient tool to locate the suitable perforator or dominant vessel providing sufficient blood supply to the flap, intraoperatively, and can also help in predicting postoperative vascularization complications (Thiessen et al., 2020). Other methods commonly used in this aspect are handheld Doppler, color Doppler ultrasound, magnetic resonance angiography, and computed tomography angiography, but DIRT seems to be a valuable investigation for the pre-, per-, and postoperative phases of DIEP-flap reconstructions.

8.9.3 Renal surgery

Inducing temporary renal ischemia is a necessary venture in renal surgeries, particularly partial nephrectomies. Nephron sparing surgery (NSS) or better understood as partial nephrectomy, is the gold standard approach to small renal tumors in most cases. The response to this inevitable ischemic period is advantageous to the treating surgeon in terms of guiding appropriate strategies to prevent renal injury or minimize its harmful effects. Studies have managed to identify renal response to ischemia by using hyperspectral imaging, which is highly relevant in laparoscopic or robot-assisted approaches. The laparoscopic hyperspectral imaging system has the ability to identify the dynamic changes in renal oxygenation status during the procedure and predict the postoperative results arising from the ischemic and postischemic renal status (Lu & Fei, 2014).

8.9.4 Plastic surgery

HSI has the potential for noninvasive and contactless assessment of flaps used in plastic surgery along with providing spatial resolution about the flap accurately. Problems often arise from postoperative vascularization failure of tissue flaps. The intraoperative prediction regarding the viability of flaps in the postoperative period is under study, although definite results regarding the convenience of using such methods are still underway (Holmer et al., 2016).

8.9.5 Transplant surgeries

Liver transplant: Experimental methods or promising tools to monitor liver ischemia during liver transplant operations include carbon dioxide sensor, indocyanine green (ICG), and near-infrared spectroscope. However, drawbacks like invasiveness and lack of ischemic quantification are difficult for proper clinical practice. HSI technology allows the intraoperative, noninvasive solution to these problems with standardized perfusion assessment (Felli et al., 2020). Nonoperator dependence is an additional advantage, as is the optical quantification of endogenous molecules without labeled dye injection. In this context, it is relevant to speak about liver ischemia and reperfusion injury (IRI), a dreaded complication due to disruption of parenchymal and microvascular architecture leading to hepatic functional impairment. This is pertinent in liver transplantation and during liver surgery performed with intermittent vascular inflow occlusion. Parenchymal disruption in the reperfusion phase mainly depends on the ischemic time duration. Consequently, intraoperative localization and quantification of oxygen impairment may be helpful to quickly detect future reperfusion injury sites.

    Renal transplant: Evaluation of graft function prior to transplantation is important in view of high rates of delayed graft function or acute rejection of organs. The latter is common in cases of donors with a comorbid feature such as hypertension, diabetes mellitus, elderly aged donor, or compensated kidney function. Currently, the gold standard preservation technique is the static cold storage of the organ in ice slush or icebox, and the organ condition is assessed by the surgeon subjectively. Normothermic machine perfusion with autologous blood is an alternative, as a near-physiologic storage system, and along with it HSI imaging can prove to be a beneficial tool in assessment of the kidney functionality in a contactless, noninvasive, and rapid manner.

8.9.6 Colorectal cancer surgeries

The surgical approach remains the mainstay for inoperable and resectable colorectal cancer. This has evolved from open to the laparoscopic approach, due to the better postoperative period and a similar outcome. However, the main drawback in laparoscopic approach remains the absence of haptic assessment on the part of the surgeon, which can compromise the efficacy of resection. HSI has the ability in these scenarios to help the surgeon intraoperatively demarcate healthy colorectal margin from tumor tissue without tactile feedback. Ex vivo automatic classification of colonic biopsy samples has been proposed based on HSI (Masood & Rajpoot, 2009).

8.10 Applications for imaging the eye
8.10.1 Retinal imaging

HSI has been strongly advocated for diagnosis of various retinal diseases. Diabetic retinopathy is a widespread condition and needs early diagnosis for appropriate intervention. In addition to different scanning technologies available for diabetic retinopathy, HSI can reduce the time to diagnosis and, with the help of diagnostic heuristics, can reduce operator error (Cole et al., 2016). It can also help in the diagnosis of age-related macular degeneration (ARMD) or radiation retinopathy (Rose et al., 2018; Tong et al., 2016). The diagnoses may not be limited to the eye but also be helpful to diagnose diseases of the central nervous system like Alzheimer disease (Hadoux et al., 2019).

8.11 In vitro imaging
8.11.1 Models of wound healing

In regenerative medicine, it is imperative to have representative models for wound healing and imaging techniques to quantify epithelialization. The information from in vitro wound healing models can be amplified using HSI (Bjorgan et al., 2020). With the advent of 3D printing, people are developing artificial organs such as pancreases based on 3D scaffolds. HSI may have a role in accessing the successful engraftment of such 3D scaffolds by induced stem cells to develop into pancreatic cells.

8.11.2 HSI-based light microscopy

Modern-day in vitro imaging uses different fluorochromes to label various cells, receptors, or ligands. Numerous filters and cameras are required to image such systems. Confocal microscopy is often too slow at this (relatively large) scale. Development of an HSI system can simplify this process as well as avoid phototoxicities of such complicated systems (Jahr et al., 2015).

8.11.3 Confocal microscopy

Confocal microscopy with HSI features has been developed more than a decade ago (Sinclair et al., 2006). Currently phenotypically similar-looking but functionally different cells such as M1 versus M2 macrophages can be differentiated by an HSI integrated confocal microscopy system (Bertani et al., 2017). This has great potential because it is noninvasive and does not disturb the natural interactions between cells with the introduction of various foreign proteins like fluorochromes.

8.12 Conclusion

With the advent of modern sensors, solid-state devices, and the jump in computational ability supporting artificial intelligence, HSI has progressed leaps and bounds. Proofs of concepts are available in various biomedical fields from noninvasive diagnosis to endoscopies to cancer or transplant surgeries. As the costs of these technologies are coming down, it is expected that hyperspectral imaging will come in a big way to aid clinical diagnosis and laboratory medicine in the near future.

References

Allen, J., Howell, K., 2014. Microvascular imaging: Techniques and opportunities for clinical physiological measurements. Physiological Measurement 35, R91–R141.

Beletkaia, E., Dashtbozorg, B., Jansen, R.G., Ruers, T.J.M., Offerhaus, H.L., 2020. Nonlinear multispectral imaging for tumor delineation. Journal of Biomedical Optics 25.

Bertani, F.R., Mozetic, P., Fioramonti, M., Iuliani, M., Ribelli, G., Pantano, F., Santini, D., Tonini, G., Trombetta, M., Businaro, L., Selci, S., Rainer, A., 2017. Classification of M1/M2-polarized human macrophages by label-free hyperspectral reflectance confocal microscopy and multivariate analysis. Scientific Reports 7, 8965.

Bjorgan, A., Pukstad, B.S., Randeberg, L.L., 2020. Hyperspectral characterization of re-epithelialization in an in vitro wound model. Journal of Biophotonics, e202000108.

Cole, E.D., Novais, E.A., Louzada, R.N., Waheed, N.K., 2016. Contemporary retinal imaging techniques in diabetic retinopathy: A review. Clinical & Experimental Ophthalmology 44, 289–299.

Daeschlein, G., Langner, I., Wild, T., von Podewils, S., Sicher, C., Kiefer, T., Jünger, M., 2017. Hyperspectral imaging as a novel diagnostic tool in microcirculation of wounds. Clinical Hemorheology and Microcirculation 67, 467–474.

DSouza, A.V., Lin, H., Henderson, E.R., Samkoe, K.S., Pogue, B.W., 2016. Review of fluorescence guided surgery systems: Identification of key performance capabilities beyond indocyanine green imaging. Journal of Biomedical Optics 21, 80901.

Fabelo, H., Ortega, S., Lazcano, R., Madroñal, D., M. Callicó, G., Juárez, E., Salvador, R., Bulters, D., Bulstrode, H., Szolna, A., Piñeiro, J.F., Sosa, C., J. O'Shanahan, A., Bisshopp, S., Hernández, M., Morera, J., Ravi, D., Kiran, B.R., Vega, A., … Sarmiento, R., 2018. An intraoperative visualization system using hyperspectral imaging to aid in brain tumor delineation. Sensors, 18.

Fauvel, M., Tarabalka, Y., Benediktsson, J.A., Chanussot, J., Tilton, J.C., 2013. Advances in spectral-spatial classification of hyperspectral images. Proceedings of IEEE 101, 652–675.

Felli, E., Al-Taher, M., Collins, T., Baiocchini, A., Felli, E., Barberio, M., Ettorre, G.M., Mutter, D., Lindner, V., Hostettler, A., Gioux, S., Schuster, C., Marescaux, J., Diana, M., 2020. Hyperspectral evaluation of hepatic oxygenation in a model of total vs. arterial liver ischaemia. Scientific Reports 10, 15441.

Gioux, S., Choi, H.S., Frangioni, J.V., 2010. Image-guided surgery using invisible near-infrared light: Fundamentals of clinical translation. Molecular Imaging 9, 237–255.

Hadoux, X., Hui, F., Lim, J.K.H., Masters, C.L., Pébay, A., Chevalier, S., Ha, J., Loi, S., Fowler, C.J., Rowe, C., Villemagne, V.L., Taylor, E.N., Fluke, C., Soucy, J.-P., Lesage, F., Sylvestre, J.-P., Rosa-Neto, P., Mathotaarachchi, S., Gauthier, S., … van Wijngaarden, P., 2019. Non-invasive in vivo hyperspectral imaging of the retina for potential biomarker use in Alzheimer's disease. Nature Communications 10, 4227.

Halicek, M., Lu, G., Little, J.V., Wang, X., Patel, M., Griffith, C.C., El-Deiry, M.W., Chen, A.Y., Fei, B., 2017. Deep convolutional neural networks for classifying head and neck cancer using hyperspectral imaging. Journal of Biomedical Optics 22, 60503.

Heinz, D.C., Chang, C.-I., 2001. Fully constrained least squares linear spectral mixture analysis method for material quantification in hyperspectral imagery. IEEE Transactions on Geoscience and Remote Sensing 39, 529–545.

Hirano, G., Nemoto, M., Kimura, Y., Kiyohara, Y., Koga, H., Yamazaki, N., Christensen, G., Ingvar, C., Nielsen, K., Nakamura, A., Sota, T., Nagaoka, T., 2020. Automatic diagnosis of melanoma using hyperspectral data and GoogleNet. Skin Research and Technology 26, 891–897.

Hohmann, M., Kanawade, R., Klämpfl, F., Douplik, A., Mudter, J., Neurath, M.F., Albrecht, H., 2017. In-vivo multispectral video endoscopy towards in-vivo hyperspectral video endoscopy. Journal of Biophotonics 10, 553–564.

Holmer, A., Tetschke, F., Marotz, J., Malberg, H., Markgraf, W., Thiele, C., Kulcke, A., 2016. Oxygenation and perfusion monitoring with a hyperspectral camera system for chemical based tissue analysis of skin and organs. Physiological Measurement 37, 2064–2078.

Jahr, W., Schmid, B., Schmied, C., Fahrbach, F.O., Huisken, J., 2015. Hyperspectral light sheet microscopy. Nature Communications 6, 7990.

Jansen-Winkeln, B., Maktabi, M., Takoh, J.P., Rabe, S.M., Barberio, M., Köhler, H., Neumuth, T., Melzer, A., Chalopin, C., Gockel, I., 2018. [Hyperspectral imaging of gastrointestinal anastomoses]. Der Chirurg 89, 717–725.

Jen, C., Huang, C., Chen, Y., Kuo, C., Wang, H., 2014. Diagnosis of human bladder cancer cells at different stages using multispectral imaging microscopy. IEEE Journal of Selected Topics in Quantum Electronics 20, 81–88.

Kiyotoki, S., Nishikawa, J., Okamoto, T., Hamabe, K., Saito, M., Goto, A., Fujita, Y., Hamamoto, Y., Takeuchi, Y., Satori, S., Sakaida, I., 2013. New method for detection of gastric cancer by hyperspectral imaging: A pilot study. Journal of Biomedical Optics 18, 26010.

Kriegmair, M.C., Rother, J., Grychtol, B., Theuring, M., Ritter, M., Günes, C., Michel, M.S., Deliolanis, N.C., Bolenz, C., 2020. Multiparametric cystoscopy for detection of bladder cancer using real-time multispectral imaging. European Urology 77, 251–259.

Kumashiro, R., Konishi, K., Chiba, T., Akahoshi, T., Nakamura, S., Murata, M., Tomikawa, M., Matsumoto, T., Maehara, Y., Hashizume, M., 2016. Integrated endoscopic system based on optical imaging and hyperspectral data analysis for colorectal cancer detection. Anticancer Research 36, 3925–3932.

Leon, R., Martinez-Vega, B., Fabelo, H., Ortega, S., Melian, V., Castaño, I., Carretero, G., Almeida, P., Garcia, A., Quevedo, E., Hernandez, J.A., Clavo, B., M. Callico, G., 2020. Non-invasive skin cancer diagnosis using hyperspectral imaging for in-situ clinical support. Journal of Clinical Medicine, 9.

Liu, N., Chen, Z., Xing, D., 2020. Integrated photoacoustic and hyperspectral dual-modality microscopy for co-imaging of melanoma and cutaneous squamous cell carcinoma in vivo. Journal of Biophotonics 13, e202000105.

Livingston, E.H., Gulaka, P., Kommera, S., Wang, B., Liu, H., 2009. In vivo spectroscopic characterization of porcine biliary tract tissues: First step in the development of new biliary tract imaging devices. Annals of Biomedical Engineering 37, 201–209.

Lu, G., Fei, B., 2014. Medical hyperspectral imaging: A review. Journal of Biomedical Optics 19, 010901.

Ma, L., Lu, G., Wang, D., Wang, X., Chen, Z.G., Muller, S., Chen, A., Fei, B., 2017. Deep learning based classification for head and neck cancer detection with hyperspectral imaging in an animal model, Proceedings of SPIE International Society for Optical Engineering, 10137.

Marotz, J., Kulcke, A., Siemers, F., Cruz, D., Aljowder, A., Promny, D., Daeschlein, G., Wild, T., 2019. Extended perfusion parameter estimation from hyperspectral imaging data for bedside diagnostic in medicine. Molecules, 24.

Masood, K., Rajpoot, N., 2009. Texture based classification of hyperspectral colon biopsy samples using CLBP, 2009 IEEE International Symposium on Biomedical Imaging: From Nano to Macro (pp. 1011–1014). Presented at the 2009 IEEE International Symposium on Biomedical Imaging: From Nano to Macro (ISBI). IEEE, Boston, MA, USA.

Milanic, M., Paluchowski, L.A., Randeberg, L.L., 2015. Hyperspectral imaging for detection of arthritis: Feasibility and prospects. Journal of Biomedical Optics 20, 096011.

Monteiro, S. T., Uto, K., Kosugi, Y., Oda, K., Iino, Y., & Saito, G. (n.d.). Hyperspectral image classification of grass species in Northeast Japan (p. 2).

Panasyuk, S.V., Yang, S., Faller, D.V., Ngo, D., Lew, R.A., Freeman, J.E., Rogers, A.E., 2007. Medical hyperspectral imaging to facilitate residual tumor identification during surgery. Cancer Biology and Therapy 6, 439–446.

Pierce, M.C., Schwarz, R.A., Bhattar, V.S., Mondrik, S., Williams, M.D., Lee, J.J., Richards-Kortum, R., Gillenwater, A.M., 2012. Accuracy of in vivo multimodal optical imaging for detection of oral neoplasia. Cancer Prevention Research (Philadelphia. PA) 5, 801–809.

Plaza, A., Benediktsson, J.A., Boardman, J.W., Brazile, J., Bruzzone, L., Camps-Valls, G., Chanussot, J., Fauvel, M., Gamba, P., Gualtieri, A., Marconcini, M., Tilton, J.C., Trianni, G., 2009. Recent advances in techniques for hyperspectral image processing. Remote Sensing of Environment 113, S110–S122.

Predina, J.D., Newton, A., Corbett, C., Xia, L., Sulyok, L.F., Shin, M., Deshpande, C., Litzky, L., Barbosa, E., Low, P.S., Kucharczuk, J.C., Singhal, S., 2018. Localization of pulmonary ground-glass opacities with folate receptor–targeted intraoperative molecular imaging. Journal of Thoracic Oncology 13, 1028–1036.

Rose, K., Krema, H., Durairaj, P., Dangboon, W., Chavez, Y., Kulasekara, S.I., Hudson, C., 2018. Retinal perfusion changes in radiation retinopathy. Acta Ophthalmology (Copenhagen) 96, e727–e731.

Sicher, C., Rutkowski, R., Lutze, S., von Podewils, S., Wild, T., Kretching, M., Daeschlein, G., 2018. Hyperspectral imaging as a possible tool for visualization of changes in hemoglobin oxygenation in patients with deficient hemodynamics—proof of concept. Biomed Tech (Berl) 63, 609–616.

Sinclair, M.B., Haaland, D.M., Timlin, J.A., Jones, H.D.T., 2006. Hyperspectral confocal microscope. Applied Optics 45, 6283–6291.

Thiessen, F.E.F., Vermeersch, N., Tondu, T., Van Thielen, J., Vrints, I., Berzenji, L., Verhoeven, V., Hubens, G., Verstockt, J., Steenackers, G., Tjalma, W.A.A., 2020. Dynamic infrared thermography (DIRT) in DIEP flap breast reconstruction: A clinical study with a standardized measurement setup. European Journal of Obstetrics & Gynecology and Reproductive Biology 252, 166–173.

Toivonen, M.E., Rajani, C., Klami, A., 2021. Snapshot hyperspectral imaging using wide dilation networks. Machine Vision and Applications, 32.

Tong, Y., Ben Ami, T., Hong, S., Heintzmann, R., Gerig, G., Ablonczy, Z., Curcio, C.A., Ach, T., Smith, R.T., 2016. Hyperspectral Autofluorescence Imaging of drusen and retinal pigment epithelium in donor eyes with age-related macular degeneration. Retina (Philadelphia, PA) 36 (1), S127–S136.

Ujiie, H., Kato, T., Hu, H., Patel, P., Wada, H., Fujino, K., Weersink, R., Nguyen, E., Cypel, M., Pierre, A., de Perrot, M., Darling, G., Waddell, T.K., Keshavjee, S., Yasufuku, K., 2017. A novel minimally invasive near-infrared thoracoscopic localization technique of small pulmonary nodules: A phase I feasibility trial. Journal of Thoracic and Cardiovascular Surgery 154, 702–711.

Wu, I.-C., Syu, H.-Y., Jen, C.-P., Lu, M.-Y., Chen, Y.-T., Wu, M.-T., Kuo, C.-T., Tsai, Y.-Y., Wang, H.-C., 2018. Early identification of esophageal squamous neoplasm by hyperspectral endoscopic imaging. Scientific Reports 8, 13797.

Ye, F., Li, M., Zhu, S., Zhao, Q., Zhong, J., 2018. Diagnosis of dermatophytosis using single fungus endogenous fluorescence spectrometry. Biomedical Optics Express 9, 2733–2742.

Yoon, J., Joseph, J., Waterhouse, D.J., Luthman, A.S., Gordon, G.S.D., di Pietro, M., Januszewicz, W., Fitzgerald, R.C., Bohndiek, S.E., 2019. A clinically translatable hyperspectral endoscopy (HySE) system for imaging the gastrointestinal tract. Nature Communications, 10.

Zimmermann, P., Curtis, N., 2019. Factors that influence the immune response to vaccination. Clinical Microbiology Reviews, 32.

Zuzak, K.J., Naik, S.C., Alexandrakis, G., Hawkins, D., Behbehani, K., Livingston, E.H., 2007. Characterization of a near-infrared laparoscopic hyperspectral imaging system for minimally invasive surgery. Analytical Chemistry 79, 4709–4715.

PA Imaging: A promising tool for targeted therapeutic implications in Cancer

9

Samudyata C. Prabhuswamimath

Unit for Human Genetics, All India Institute of Speech and Hearing, Manasagangothri, Mysore, India

9.1 Introduction

Medical imaging has revolutionized clinical diagnosis and therapeutics through non-invasive technologies. It has made possible the exploration of physiological processes within living systems with enhanced sensitivity and specificity, resulting in better diagnostic and treatment regimens (Ding et al., 2018). Some of the advantages of optical imaging are the use of nonionizing radiation, cost effectiveness, and accessibility to various optical contrast agents. However, due to strong light scattering, it produces shallow penetration and low resolution. Hence photoacoustic (PA) imaging was devised that has the advantage of optical absorption contrast and enhanced ultrasonic resolution (Song & Wang, 2007). In general, PA imaging is also called optoacoustic imaging that combines the advantages of high-contrast, spectroscopic- based specificity of optical imaging with the high spatial resolution offered by ultrasound imaging. This specificity due to greater penetration depths than other optical imaging technologies makes it a preferred imaging technology that can detect hemoglobin, lipids, water, and light-absorbing chromophores, and assists anatomical studies of microvasculature, and functional studies like oxygenation of the blood, blood flow, temperature, and molecular imaging without artifacts. This information can be obtained over a wide range of length scales from micrometers to centimeters with scalable spatial resolution. With the appropriate enhancement of optical contrast, the deeply embedded objects at depths of 5 cm in various biologic tissues have been successfully imaged with high resolution (Beard, 2011; Kim et al., 2010).

By fine-tuning the wavelength of the incident light, various techniques based on this effect use the optical absorption spectrum of different biologic chromophores present in living systems to produce high-resolution imaging specifically. Apart from the above-mentioned biologic targets, other clinical targets are nerves that allow comparison of mobility and molecular interactions between ion channel mediated signal transduction, vascular imaging, and surgical navigation (Graham & Bell, 2020; Kim et al., 2004). Thus PA imaging offers diverse and significant applications in clinical research and medicine, basic research of disorders like cancer, cardiovascular disease, and other circulatory disorders.

Biomedical Imaging Instrumentation. DOI: https://doi.org/10.1016/B978-0-323-85650-8.00009-7

9.1.1 **Principle**

As discussed before, in PA imaging, the ultrasound waves are generated by irradiating the biological tissue with electromagnetic radiation pulsed for nanoseconds (Fig. 9.1). However, various modulation techniques are available in practice. The most commonly used optical wavelengths in the visible and near-infrared (NIR) range from 550 to 900 nm, wherein imaging at greater depths of centimetres is done as it offers better penetration depths. However, thermoacoustic imaging uses longer wavelengths into the microwave band (300 MHz–3 GHz), leading to better penetration depths. During optical excitation, absorption by tissue chromophores is followed by rapid conversion to heat and an increase in temperature; generally, less than 0.1 K. Temperature rise in this range does not cause any physiological changes or physical damage to the biologic tissues. This phenomenon leads to increased pressure, followed by the emission of acoustic waves that are tens of megahertz with a low amplitude less than 10 kPa. These are detected by either a single or an array of mechanically scanned ultrasound receivers for the acquisition of A-lines. An image is reconstructed by measuring the time of arrival of acoustic waves and the speed of sound. It is almost similar to the way a conventional pulse-echo ultrasound image is formed. The previously acquired A-lines are used for image construction or in conjunction with reconstruction algorithms that work on back-projection or phased array beam-forming principles. In the case of image formation in pulse-echo ultrasound, localization is achieved by focusing both the transmit beam and the receive beam. Generally, the soft tissues at greater penetration depths exhibit overwhelming optical scattering. Thus, in PA imaging, localization is obtained by focusing the transmit signal—the excitation light and is hence achieved by reception only. Another distinguishing factor is the magnitude of acoustic pressures involved, wherein ultrasound scanners in a clinical diagnostic setup produce focal peak pressures of about 1 MPa, in contrast to PA pressure amplitudes in the range of 10 kPa, which is comparatively of lesser magnitude. Harmonic imaging, which exploits the nonlinear propagation of ultrasound through the living tissues produces better image quality than conventional ultrasound technology. As PA pressure amplitudes are lower in magnitude, the nonlinear acoustic propagation is not witnessed in PA imaging. This also indicates superior safety from hazardous ultrasound exposure; however, safety considerations from a laser cannot be ruled out (Beard, 2011; Lashkari & Mandelis, 2010).

It is essential to understand why different imaging techniques that have evolved over a period of time have vast applications in cancer. Also, a comprehensive understanding of the angiogenic and metastatic processes in cancer and the molecular mechanism involved in tumorigenesis is necessary to exploit the advanced vascular imaging technologies for diagnosis and therapy at a physiological and molecular level. Therefore, the following sections will provide an overview of angiogenesis emphasizing Vascular Endothelial Growth Factor (VEGF), metastasis, and critical cellular signaling pathways.

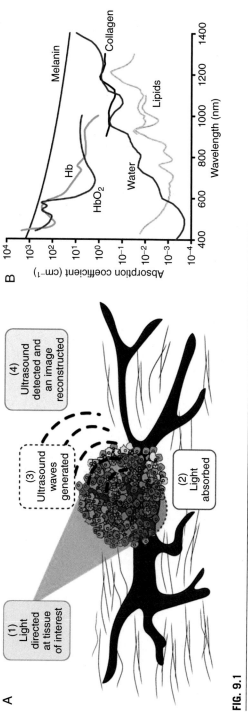

FIG. 9.1

Principles of photoacoustic imaging (PAI) https://dmm.biologists.org/content/12/7/dmm039636 (Brown et al., 2019).

9.1.2 **Breast Cancer**

According to the GLOBOCAN Cancer Tomorrow prediction tool, breast cancer is the most common malignant disease among women worldwide, and incident cases are expected to increase by more than 46% by 2040 (Bray et al., 2018). Cancer is a significant cause of morbidity and the second leading cause of mortality in the United States. Prostate, lung, and colorectal cancers account for 43% of all cases in men; breast, lung, and colorectal cancers are the most common cancers among women accounting for 50% of new diagnoses, and breast cancer accounting for 30% of female cancers. Though some hematopoietic and lymphoid malignancies have witnessed breakthrough treatments like immunotherapy and targeted therapy for metastatic melanoma, there is considerable slow progress in cancers that are amenable to early detection and screening programs like breast cancer, prostate, and colorectal cancer (Siegel et al., 2020). Among heterogeneous populations like India, the projected number of cancer patients in 2020 is 1,392,179, and breast, lung, mouth, cervix uteri, and tongue would dominate the other cancers (Mathur et al., 2020). Over the last three decades, there has been a substantial reduction in breast cancer mortality in high-income countries due to advancements in treatment, diagnosis, and disease management. However, persistent disparities in the global decline in breast cancer emphasize immediate attention, especially in middle-income countries (Carioli et al., 2018). There are various types of classifications of tumors. The ICD 10 (International Classification of Diseases, 10th Edition) classifies breast malignancy at C50. The World Health Organization has defined the standards for classification and nomenclature for pathologists worldwide (Sinn & Kreipe, 2013). The staging of breast cancers is based on surface receptors, tissue, node, and metastasis (Mittendorf et al., 2018). The prognosis of breast cancer is influenced by the stage of cancer (van Hellemond et al., 2018) and hormonal receptors (Cubasch et al., 2018). It cannot be ruled out that comorbidities and ethnicity also play an important role (Ewertz et al., 2018). This hormonal cancer of the mammary gland has distinct molecular causes and is influenced by the premenopausal and postmenopausal condition of women. A recent interesting study shows that approximately 645,000 premenopausal and 1.4 million postmenopausal breast cancer cases were reported globally in 2018 leading to 130,000 and 490,000 deaths in each group, respectively (Heer et al., 2020). Early detection and timely therapeutic intervention increase the survival rate of breast cancer patients. Significant developments in imaging technologies, from x-ray imaging to mammography, have accelerated the early detection and therapy (Aydın & Torun, 2020). Mammography aids in the detection of breast cancer at earlier stages with a majority of successful treatment outcomes. However, it is affected by false positives and false negatives. The advancement of technology in diagnostics has led to the introduction of an artificial intelligence (AI) system that provides precision for breast cancer predictions (McKinney et al., 2020). X-ray mammography and ultrasonography focus on the morphologic changes associated with breast tissue but are not very accurate in detection. A detection system based on tumor angiogenesis focusing on microvessel density and aberrant vascular characteristics produces sensitive and specific results. Despite being popular, dynamic contrast-enhanced magnetic resonance imaging (MRI) is sensitive but not specific and inefficient with the detection of ductal carcinoma in situ (DCIS). Techniques like contrast-enhanced digital mammography can detect DCIS but is dependent on hazardous ionizing radiation and raises health concerns. Despite detecting intravascular blood volume and flow, contrast-enhanced ultrasound has difficulties in differentiating benign from malignant tissue. All of these imaging modalities are based on injections of contrast agents, whereas newer modalities like diffuse optical imaging and photoacoustic imaging do not require external contrast agents. As they both detect a high concentration of hemoglobin in malignancy, high intrinsic contrast is generated (Heijblom et al., 2011).

9.2 **Angiogenesis**

The construction of blood vessels takes place by two processes, vasculogenesis and angiogenesis. The differentiation of progenitor endothelial cells to endothelial cells leads to vasculogenesis. The capillary sprouting of new vessels from pre-existing small vessels is called angiogenesis (Fig. 9.2) (Risau, 1997). Angiogenesis occurs in normal physiologic cases like restoration of blood supply to facilitate wound healing, menstruation, growth, and development. However, it plays a vital role in tumor progression and metastasis (Lou et al., 2017; Todorova et al., 2017). The tumor microenvironment is composed of both tumor and stromal cells, extracellular matrix, and secretory factors in general. The changes in gene expression lead to the disruption of tissue homeostasis and overexpression and secretion of cytokines and growth factors, among others (Zuazo-Gaztelu & Casanovas, 2018). During tumorigenesis,

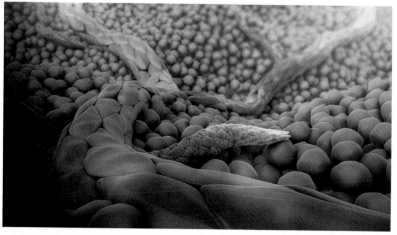

FIG. 9.2

Process of Angiogenesis From Wikimedia Commons, http://www.scientificanimations.com/wiki-images/.

the positive regulators of angiogenesis are overexpressed, and angiogenic proteins get mobilized from the extracellular matrix. The recruitment of macrophages takes place leading to the release of angiogenic proteins, the most prominent ones being basic fibroblast growth factor (bFGF) and vascular endothelial growth factor (also known as vascular permeability factor), which are key players in angiogenesis (Folkman, 1995). Anti-angiogenic therapy is garnering widespread interest by the fact that appropriate anti-angiogenic therapeutic modalities can convert tumor microenvironment (TME) from immunosuppressive to immunosupportive status (Yi et al., 2019). Despite the great potential immunotherapy offers in oncology, vascular abnormalities play a significant role in immune evasion due to the release of proangiogenic factors like VEGF and angiopoietin 2 (ANG2). The major immune checkpoint molecules are programmed cell death protein 1 (PD-1) and cytotoxic T-lymphocyte antigen-4 (CTLA-4) that downregulate the magnitude of an immune response in homeostatic conditions. However, upregulation of PD-1/PD-L1–mediated immune signaling pathways promotes cancer cells to evade immune surveillance (Yi et al., 2018). This makes such immune checkpoint molecules and their ligands most suitable for targeted therapy in cancer. Recent understanding that the potential of antiangiogenic therapy is not limited to inhibiting tumor progression and metastasis but also reprogram the tumor immune microenvironment has encouraged many preclinical studies on combinatorial therapy of immune checkpoint inhibitor (ICI) and antiangiogenic therapy. The results have demonstrated superior efficacy compared to monotherapy and deployed clinical trials with promising results (Yi et al., 2019). Immunotherapeutic approaches rely on the activity of immune effector cells within the TME. Hence, a combinatorial approach of immunotherapy, and antiangiogenic therapy specifically targeting key players like VEGF and ANG2 would give better results and improved patient outcomes (Fukumura et al., 2018).

In the following sections, we will understand the role of VEGF as a proangiogenic and permeability-inducing factor and its effect on tumorigenesis and metastasis.

9.2.1 Vascular Endothelial Growth Factor

VEGF was isolated and cloned in 1989, and it was a breakthrough year that paved the way for understanding the mechanism of angiogenesis in different contexts (Ferrara & Adamis, 2016). Angiogenesis is promoted by a variety of growth factors like FGF, bFGF (basic Fibroblast Growth Factor), VEGF, MMPs (matrix metalloproteinases), and angiopoietins. However, the VEGF-mediated signaling pathway is the most crucial and primary stimuli for angiogenic processes that are aberrant in the tumor context. This leads to the continuous blood supply and propagation of tumor cells. Mechanistically, VEGF acts as a promoter for proliferation, migration, and differentiation of endothelial cells that converge for blood vessel formation and blood flow. VEGF belongs to the platelet-derived growth factor family with five members, namely VEGF-A, VEGF-B, VEGF-C, VEGF-D, and P1GF (placental growth factor) in mammalian genomes including humans. VEGF-A regulates angiogenesis and permeability by the activation of two receptors, namely VEGFR-1 (Flt-1) and

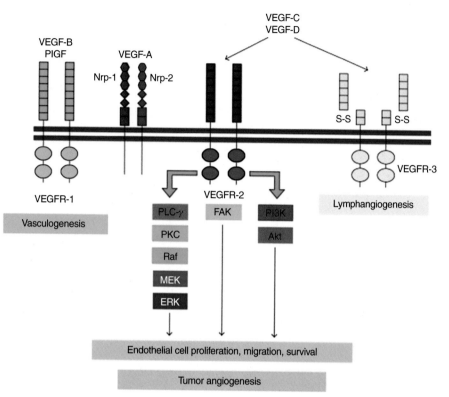

FIG. 9.3

VEGF Signaling Pathway https://www.researchgate.net/figure/VEGF-signaling-interactions-The-VEGF-family-can-bind-to-VEGFR1-VEGFR2-and-VEGFR3_fig1_45185768 (Korpanty et al., 2010).

VEGFR-2 (KDR/Flk1 in mice) (Fig. 9.3). VEGF-C, VEGF-D, and their receptor, VEGFR-3, are involved in the regulation of lymphangiogenesis. However, the major signal transduction pathway in angiogenesis PLCγ-PKC-MAPK is mediated by VEGFR-2 (Nishida et al., 2006; Shibuya, 2011). Recent advances in angiogenesis research have proved that VEGF-A plays a key role in regulating angiogenesis in tissue homeostasis and pathologic context (Apte et al., 2019).

9.2.2 VEGF and Permeability

This section discusses the role of VEGF as a permeability-inducing factor and its implications on tumor biology and treatment outcome.

The VEGF/vascular permeability factor secreted by tumor cells is a 34 to 42 kDa disulfide-bonded glycoprotein that is a dimeric and heparin-binding protein. It acts on endothelial cells by receptor activation of phospholipase C and induction of [Ca2+]i transients. Due to its potent permeability property, it promotes the extravasation of plasma fibrinogen and its deposition causing alterations in the extracellular matrix (ECM). This altered matrix facilitates the ingrowth of macrophages, fibroblasts, and endothelial cells leading to endothelial cell proliferation. It is also involved in the induction of malignant ascites (Senger et al., 1993).

The blockage of VEGF inhibited hyperpermeability, and this potentially transformed its clinical applications in oncology. As stated, the crosslinking of fibrin fibers is promoted by the enzyme transglutaminase, thereby promoting tumor angiogenesis and wound healing (Haroon et al., 2002). This leads to the activation of transforming growth factor-β (TGFβ) that promotes collagen deposition and fibrosis. Though fibrosis assists in physiological conditions like wound healing, it negatively influences tumor context by increasing the ECM extensively depending on the type of cancer, thereby leading to tumor cell movement and metastasis. These alterations in ECM act as a barrier for drug transport affecting the diffusion of drugs from the vasculature to the target tumor cells (Minchinton & Tannock, 2006) especially in the case of nanoparticles (NPs) as drug delivery systems (Miao & Huang, 2015). Hyperpermeability also poses the challenge of elevated interstitial fluid pressure (IFP) in solid tumors as a consequence of VEGF-mediated hyperpermeability. The lymphatic vasculature is significant in maintaining fluid balance in tissues deficient in tumors. A delicate balance between the pressure gradient and concentration gradient across the vascular wall is vital for drug transport. For macromolecular drugs, elevated IFP negatively influences the transport of macromolecular drugs and is associated with low survival in cervical cancer patients undergoing radiotherapy and chemoradiotherapy. Recent studies have shown that IFP increases tumor cell aggression and promotes epithelial-mesenchymal transition mediated metastasis (Dewhirst & Ashcraft, 2016). This implies the direct and indirect role of VEGF as a permeability-inducing factor in tumorigenesis and drug delivery systems. However, drugs targeting vessel hyperpermeability have emerged, like oral etoposide that inhibits endothelial and tumor cell proliferation and VEGF-induced angiogenesis and permeability (Panigrahy et al., 2010).

9.2.3 VEGF expression

VEGF family expression has been extensively studied in different cancers like oral cancer, gingival cancer, esophageal cancer, gastric cancer, colorectal cancer, liver cancer, and gallbladder cancer that emphasize VEGF and its receptors as biomarkers and prognostic factors for tumor metastasis (Costache et al., 2015). Elevated expressions of VEGFR-1 and VEGF were witnessed in breast cancer, especially VEGF overexpression in lymph node–positive breast cancer patients (Srabovic et al., 2013). VEGF overexpression is seen in head and neck cancer (Andisheh-Tadbir et al., 2014; Zang et al., 2013). VEGF is implicated in the malignant progression of non-small cell lung cancer (NSCLC) (Jiang et al., 2014) and ovarian cancer (Zhang et al., 2014). VEGF is overexpressed in tumor stromal cells and different cancer cells like renal cancer, lung cancer, breast cancer, and ovarian cancer and is mediated by Ras oncogene. The binding of VEGF on VEGFR1 or VEGFR2 leads to the activation of VEGFR2, triggering the angiogenic pathway. VEGF expression is also triggered by hypoxia that is seen during tumor progression. This results when hypoxia-inducible factor 1 (HIF-1) binds to hypoxia response element (HRE) within the VEGF promoter. Inflammatory mediators like interleukin 1 (IL1), TGFβ, and prostaglandin E2 (PGE2) or mechanical stress also contribute to the stimulation of VEGF expression (Yoo and Kwon, 2013).

9.2.4 Angiopoietins

In addition to VEGF-mediated angiogenesis, the four angiopoietin ligands (Ang 1–4) play a crucial role through their receptor tyrosine kinases Tie-1 and Tie-2 expressed on vascular endothelial cells and macrophages involved in angiogenesis. Therefore, the Ang-Tie ligand-receptor system, therefore, plays a regulatory role in maintaining vascular integrity and quiescence and regulates pathological processes, including inflammation. It has been reported that Ang-1 induces vessel maturation and Ang-2 exhibits a cooperative effect with other angiogenic factors in the sprouting of angiogenesis (Fagiani & Christofori, 2013; Fiedler & Augustin, 2006). Despite the initial reports of VEGF and Ang signaling operating as mutually exclusive proangiogenic mechanisms, a recent update explains Tie-2 regulated VEGF-mediated signaling via crosstalk between the two signaling pathways. Mechanistically, VEGF ligands phosphorylate Tie-2 through Tie-1. The inhibition of Tie-2 as a result of downstream signaling pathways inhibits breast cancer growth and metastasis in vivo. This happens by the decreased availability of endothelial cells and blockade of tumor cell intravasation by macrophage cells, which is crucial for metastasis of tumor (Harney et al., 2017; Singh et al., 2009). These findings have demonstrated evidence of the Tie-2 pathway to be potentially explored for a therapeutic approach in both endothelial and tumor cells (Shim et al., 2007).

9.3 Metastasis

Metastasis is a multistep process that involves a cascade of biological events leading to the dissemination of tumor cells from primary sites to distant sites of other organs driven by the genetic mutations and epigenetic changes occurring at a cellular level (Valastyan & Weinberg, 2011). A majority of cancer-induced mortality is due to tumor metastasis primarily influenced by the stromal tumor microenvironment. Hence, researchers are emphasizing the understanding of tumor microenvironment, the three-dimensional (3D) architecture of ECM that regulate functional properties of cancerous cells and influence drug sensitivity (Pouliot et al., 2013). To metastasize, tumor cells must invade the surrounding tissue and basement membrane to enter the bloodstream or lymphatics, which involves a crosstalk between tumor and stroma. This is also impacted by the plasticity of tumor cells and the microenvironment such as stromal and endothelial cells. Millions of tumor cells are shed into the lymphatic system or bloodstream per gram of the tumor tissue during this process. These circulatory tumor cells have opened up more insights into cancer biology (Butler & Gullino, 1975; van Zijl et al., 2011).

It is interesting to note that the retainment of the long-term tumorigenic potential of some tumors depends on a small number of malignant cells capable of self-renewal like stem cells. Such tumor-initiating cells are referred to as cancer stem cells (Pardal et al., 2003). Some hematologic malignancies, breast, and brain tumors have exhibited the presence of such cells that demonstrated transformed cellular phenotypes defined by cell surface markers observed in the original tumor giving rise to secondary tumors with very few transplanted cells. However, irrespective of the self-renewal capacity of cells in primary tumors, they must possess the capacity for tumor initiation to re-establish secondary tumors at a distant site. In totality, tumor-initiating capacity, altered cellular adhesions, cell motility, resistance to extracellular death signals, basement membrane disruption, and the ECM forms a prerequisite for successful metastasis in any tumor context (Gupta & Massagué, 2006).

The metastatic cascade of events includes (1) epithelial to mesenchymal transition, (2) dissociation or shedding of tumor cells from a primary tumor into the circulation and survival, (3) invasion into the neighboring tissue, (4) intravasation into blood and lymph vessels, (5) transport through vessels, (6) extravasation into the surrounding tissue, (7) establishment of disseminated cells at a secondary site, and (8) the outgrowth of micrometastasis and macrometastasis. It is interesting to note that some types of tumors exhibit organ-specific metastasis that operates by seed and soil concept wherein both seed (the cancer cell) and soil (factors in the organ environment) determine the organ specificity (Fig. 9.4) (Chambers et al., 2002; Geiger & Peeper, 2009).

9.3.1 Epithelial to Mesenchymal Transition (EMT)

Epithelial cells lose their inherent characteristics like the presence of specialized junctional proteins and apical-basal polarized phenotype. On the other hand, the mesenchymal cells are invariably migratory and motile as they lack adhesion complexes, apical-basal polarity, and are irregular in shape. During EMT, epithelial cells lose their characteristics to acquire mesenchymal properties, becoming migratory and invasive, characterized by the expression of epithelial and mesenchymal markers. Though EMT is seen in normal developmental processes, it is significant in tumor invasion and metastatic dissemination of cancer cells to distant sites. Precisely, there are different types of EMT: type 1 that occurs in embryo development and organogenesis; type 2 that is seen in wound healing, tissue regeneration, and organ fibrosis in response to inflammation; and type 3 in tumor progression as stated earlier (Gwak et al., 2014; Lee & Nelson, 2012).

Recent studies on the elucidation of signaling pathways for EMT discuss the mediation by TGFβ, Wnt, Hedgehog, Notch, and nuclear factor-κβ. The regulation of EMT is complex due to the crosstalk between various signaling pathways, molecules, and communication between cadherin and integrin, transcription factors (like Snail,

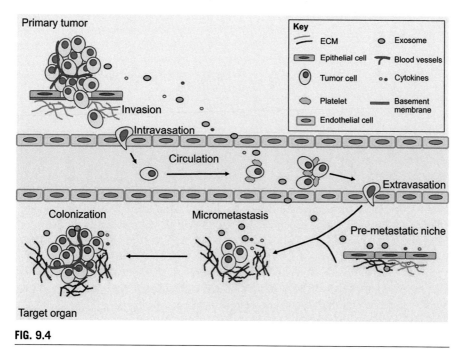

FIG. 9.4

Metastatic cascade: A multistep process https://dmm.biologists.org/content/10/9/1061 (Gómez-Cuadrado et al., 2017).

Slug, ZEB1, SIP1/ZEB2, LEF and nuclear β-catenin), growth factors like TGFβ, and PDGF. The reprogramming of epithelial cells into a more invasive mesenchymal-like tumor cell is due to the increased levels and nuclear translocation of transcription factors like Twist, Snail, Slug, ZEB1, SIP1/ZEB2, LEF, and nuclear β-catenin. Concurrently, other changes include loss of E-cadherin, increase in N-cadherin, MT1-MMP expression and secretion, and altered integrin expression profile coupled with alterations in the composition of ECM molecules (Lee & Nelson, 2012; Wu and Zhou, 2008). At the molecular level, increased levels and nuclear translocation of above-mentioned transcription factors are responsible for reprogramming epithelial cells into a more invasive, mesenchymal-like tumor cell. Concurrent events include the loss of E-cadherin, a gain of N-cadherin, increased MT1-MMP expression and MMP secretion, alterations in integrins expression profile, and changes in the composition of extracellular matrix molecule (Lee & Nelson, 2012; Wright et al., 2010).

In breast cancers, recent studies reported an upregulation of EMT markers like vimentin, smooth muscle actin, N-cadherin, and cadherin-11 and overexpression of SPARC, laminin, and fascin proteins involved in ECM remodeling and invasion. The reduction of epithelial markers like E-cadherins and cytokeratins preferentially occurred with basal-like phenotypes and expression of mesenchymal markers in epithelial components of breast carcinosarcomas (Mani et al., 2008; Sarrió et al., 2008).

The delineation of signal transduction pathways and elucidation of the molecular mechanism of EMT will lead to novel therapeutic interventions for cancer metastasis.

9.3.2 Mitogen-activated protein kinase (MAPK) pathway

MAPK pathway is an evolutionary conserved and primary signaling cascade that responds to stimuli on the cell surface, cell membrane, cytoplasm, and nucleus that is activated upon phosphorylation to affect proliferation, differentiation, self-renewal, apoptosis, angiogenesis, and metastasis in tumor context. It is primarily classified into six groups based on the source of activation: extracellular signal-regulated kinase (ERK) 1/2, ERK 3/4, ERK 5, ERK6, ERK 7/8, and Jun N-terminal kinases (JNK) 1/2/3 (Dhillon et al., 2007). The deregulation of MAPK/ERK pathway promotes tumor propagation, invasion, and metastasis by activating matrix metalloproteinases influencing the interaction of cancer cells with the ECM, and dissolution of focal adhesions (Akter et al., 2015; Joneson et al., 1996; Xie et al., 1998). VEGFR2 is the most important among the VEGF receptors, as it binds all VEGF-A isoforms, VEGF-C, and VEGF-D. Activated VEGFR2 leads to the activation of the PLC-γ-Raf kinase-MEK-MAP kinase and PI3K-AKT pathways that promote cellular proliferation and endothelial cell survival (Dang et al., 2017).

Focal adhesion kinase (FAK) is an integral signaling protein that affects proliferation, migration, angiogenesis, metastasis, self-renewal, and differentiation. It is, therefore, a key player in the metastatic cascade. Phosphorylation of FAK activates the signaling cascade via MAPK/ERK pathway and promotes various cell-specific effects. Specifically, integrin-mediated FAK activation promotes ERK1 and ERK2 signaling and promotes proliferation and cell migration (Boudreau & Jones, 1999; Gao et al., 2014). FAK has proved to be crucial for breast cancer progression and invasion and is necessary for the turnover of focal adhesions. During metastasis, the basement membrane breach occurs by localized remodeling of ECM mediated by key cellular structures called invadopodia and secretion of MMPs. The depletion of FAK induces active invadopodia formation but impairs tumor cell migration, and there is a complex interplay between FAK and Src to promote invasion in tumor context (Chan et al., 2009; Jacob & Prekeris, 2015). Interestingly, the inhibition of FAK is shown to inhibit the growth rate, proliferation, and metastatic ability of breast tumor cells in vivo (Tiede et al., 2018). These findings have encouraged researchers to identify FAK inhibitors that can be targeted for antimetastatic therapy.

With this knowledge of PA imaging and its principle, the statistics of breast cancer prevalence, the physiology of tumorigenesis and metastasis, the molecular interactions in the disease pathogenesis, and the prominent targets that are explored for therapy, we can have a better understanding of the applications of PA imaging in cancer and the significance of tumor vasculature being the most sought after parameter in modern imaging-guided therapy for cancer. The knowledge of tumor microenvironment and the properties of ECM and basement membrane provides a better understanding for clinicians and radiologists to use and modify PA imaging that offers improved imaging prospects with deeper penetration in tissues. The development and optimization of any drug targeting technology would require a complete understanding of the biological barriers and immunological surveillance as each tumor is unique in its interaction with external agents and also in its own systemic, anatomic, and cellular characteristics. Currently, multimodality imaging based on NPs has garnered extensive attention. These advances herald a new age for hybrid imaging technology in cancer diagnostics and therapeutics.

9.4 **Nanomedicine**

PA imaging has demonstrated great potential in clinical applications in recent days. However, the combination of nanotechnology and PA imaging has revolutionized translation medicine (Jiang & Pu, 2017). NPs offer high stability, high payload, multifunctionality, and are flexible for modifications. Functionally, they are efficient drug delivery systems with sustained delivery and release of drugs that make them suitable for targeted therapy. Photonics and optical techniques with biomedical applications (biophotonics) have dominated the diagnostic and therapeutics by using a wide range of light from ultraviolet (UV) to NIR (Son et al., 2019). Doxil and Abraxane are some of the Food and Drug Administration (FDA)–approved formulations for the delivery of chemotherapeutic agents like doxorubicin and paclitaxel, respectively (Anselmo & Mitragotri, 2016). Though endogenous molecules like water, lipid, hemoglobin, and melanin can absorb electromagnetic energy and produce acoustic signals without the need for contrast agents, the use of exogenous probes enhances the PA signals on the tissues, and research groups have identified hence many fluorophores, gold NPs, and carbon NPs have been identified by research groups as PA contrast agents (Lalwani et al., 2013; Weber et al., 2016).

The size of NPs is a crucial parameter with a significant impact on targeted drug delivery, biodistribution, and the amplitude of PA signals. The effect of particle size and the surface coating was investigated using d-α-tocopheryl polyethylene glycol 1000 succinates (vitamin E TPGS or TPGS). The cellular uptake of polymeric NPs for drug delivery across the gastrointestinal barrier for oral chemotherapy and blood-brain barrier for imaging and therapy of brain cancer was studied, indicating that the biodistribution of NPs was affected by particle size and surface coating. Yang et al. (2017) reported that PA intensity of perylenediimide-based semiconducting NPs increased proportionally with an increase in size from 30 nm to 200 nm that were used in PA imaging of lymph node mapping for cancer diagnosis; the 60-nm NPs exhibited optimum targeting in vivo in U87MG tumor-bearing mice (Kulkarni & Feng, 2013; Yang et al., 2017).

Some of the first generation clinically approved nanomedicines for cancer therapy are as follows: Doxil (Janssen Biotech Inc., Horsham, PA, USA), Myocet (Sopherion Therapeutics Inc., Princeton, NJ, USA), DaunoXome (Galen US Inc., Souderton, PA, USA), Depocyt (Pacira Pharmaceuticals Inc., San Diego, CA, USA), Abraxane (Celgene Corporation, Inc., Berkeley Heights, NJ, USA), Genexol-PM (Samyang Biopharmaceuticals Corporation, Jongno-gu, Seoul, Korea), and Oncaspar (Enzon

Pharmaceuticals Inc., Bridgewater, NJ, USA) (Sanna et al., 2014). The phase II clinical trial results of a programmable nanotherapeutic formulation of docetaxel called BIND-014 has revealed that it is active and well tolerated in patients with metastatic castration-resistant prostate cancer (mCRPC) undergoing chemotherapy (Autio et al., 2018).

The delivery and sustenance of NPs are influenced by the physiology of the human body that presents an acidic environment, immune surveillance, and protective mucosal linings (Roger et al., 2010). However, it is important to understand the biologic barriers like biodistribution, clearance, organ and cellular level barriers for efficient nanoparticle delivery for therapeutic applications in cancer. NPs larger than 100 nm are ideal for inhalation and intravenous administration for the treatment of lung cancers. The localization of NPs is also influenced by the pore size of endothelial and basal membranes. It is also dependent on the anatomy wherein the blood vessels in the bone are characterized by discontinuous basal membranes and wider gaps between endothelial cells that facilitate higher accumulation of nanoparticles. In contrast, it is the opposite in lungs and endocrine glands, resulting in a lower accumulation of NPs. However, the most active barrier for NP delivery would be the immune surveillance in circulation, especially the macrophages of the reticuloendothelial system (RES) that eradicate them from circulation (Gonda et al., 2019; Gustafson et al., 2015).

It is important to note that tumor vasculature and leaky vessels presented by tumors affect the distribution and delivery of nanoparticles. This enhances the accumulation of NPs in tumors leading to the enhanced permeability and retention (EPR) effect. As discussed earlier, solid tumors have defective blood vessels and exhibit high vascular permeability that facilitates the continuous supply of oxygen and nutrients for tumor growth. EPR effect exploits this anatomical-pathophysiological nature of tumor blood vessels that eliminates large macromolecules of 40 kDa or more from vessels and accumulates in tumor tissues that do not happen in a normal context. This phenomenon is explored for tumor-targeted therapy using chemotherapy and nanomedicine like Doxil, a PEGylated (polyethylene glycol-coated) liposome-encapsulated formulation of doxorubicin for the treatment of Kaposi sarcoma and other cancers (Fang et al., 2011). Furthermore, the in vivo pharmacokinetics with prolonged plasma half-life, enhanced tumor selectivity, and lesser side effects in comparison with conventional chemotherapy makes this a suitable approach for the development of other polymeric or micellar drugs, of which some are already in phase I and II trials (Duncan, 2003; Fang et al., 2011; Vicent et al., 2009).

9.4.1 Gold nanomaterials

Gold nanoparticles (AuNPs) are stimulated by NIR radiation and generate PA waves for photoacoustic imaging. They exhibit low light scattering, strong penetration, and not prone to photobleaching, making it a preferred delivery system (Rostro-Kohanloo et al., 2009). Localized surface plasmon resonance (LSPR) is one area of research based on the optical properties of nanostructures (i.e., when a metal nanostructure is illuminated with light, the photons will be absorbed or scattered at a specific wavelength) (Kreibig &

Vollmer, 2013). A novel class of nanocages has been developed by researchers based on this property that has precise tuning for LSPR and large absorptions per volume that plays a significant role in photothermal conversion (Hu et al., 2006; Skrabalak et al., 2007; Sun & Xia, 2002).

Among the numerous formulations developed to date, gold nanoshells (AuNSs) have proven to be effective theranostic agents as well as carriers. They are AuNPs made up of silica cores with gold-coated shells outside. They also display surface plasmon resonance and offer optical resonance that can be tuned to NIR regions, endowing them with thermal and mechanical effects. Also, they are biocompatible and are modifiable due to their well-established chemistry (Zhao et al., 2014). The Sprague Dawley rats' cerebral cortexes were injected with PEGylated NSs and photoacoustic computed tomography (PACT) images were captured. This proved the enhancement of contrast offered by AuNSs in PA imaging in vivo. They have been extensively applied in tumor vasculature imaging, photothermal therapy, and chemotherapy as drug carriers (Li et al., 2009; O'Neal et al., 2004).

Pan et al. (2009, 2011) have demonstrated interesting findings with gold nanobeacons that are synthesized by the entrapment of small spheric AuNPs within a colloidal particle. It is then encapsulated with a biocompatible phospholipid coating. Apart from the advantages of enhanced PA, AuNPs are easily metabolized and excreted through the urinary system. GNB1, the first-generation gold nanobeacons, exhibited stability in terms of size and zeta potential. Interestingly, the PA signals were ten times higher when rat blood was mixed with GNB1 in comparison with pure blood. Thus GNB1 was used for imaging neovascularization to study the overexpression of $\alpha v \beta 3$-integrin on endothelial cells during nonpolarized angiogenesis producing successful visualization of angiogenic processes like neovessel tubules, sprouting, and bridges (Zou et al., 2017).

A gold nanobeacon is a ssDNA oligonucleotide with a stem-loop structure consisting of fluorophore at one end and an AuNP (quencher) at the other end of the beacon. In the absence of the complementary target (native form), the stem-loop structure is closed, forcing the fluorophore and AuNP into proximity, resulting in fluorescence quenching; whereas a complementary target to the loop sequence is capable of specifically hybridizing to the hairpin, forming a double-stranded structure that opens the beacon, which spatially parts the fluorophore and AuNP, and fluorescence is restored with its applications in gene-silencing protocols (Baptista, 2014). Recent reports on the use of nanobeacons on Kras gene expression in gastric tumors have revealed a positive outcome of 60% reduction in tumor size and a 90% reduction in tumor vascularization using anti-Kras nanobeacons. In addition, the inhibition of Kras gene expression in gastric tumors is reported to prevent lung metastasis by 80% and efficiently improves survival rates in mice (Bao et al., 2015).

9.4.2 Single-walled carbon nanotubes (SWCNTs)

The SWCNTs are another class of nanoparticles with strong optical properties suitable for photoacoustic imaging (PAI) of tumor neovasculature and photothermal therapy. Here, integrin-targeted SWCNTs were synthesized by covalently conjugating SWCNTs with arginine-glycine-aspartate (RGD) peptides for the visualization of tumor neovasculature that gave higher PA signals in tumor regions as compared to conjugated control (De La Zerda et al., 2008; Zou et al., 2017). A recent study on heat-based therapy for bladder cancer reports the intravesical administration of bladder-specific SWCNTs conjugated to Annexin V at a low dose followed by a brief treatment of near-infrared light that heats only the bound nanotubes. Upon phosphatidylserine externalization, these SWNCT annexin V conjugates selectively and specifically target tumor cells, followed by NIR light treatment causing tumor cell death. This type of modification has shown promising results in mice with orthotopic MB49 murine bladder tumors with no visible tumors post 24 hours of treatment with NIR and also in other orthotopic tumors with no side effects on the healthy bladder and complete clearance of nanotubes from organs and bladder post appropriate treatment regimen (Virani et al., 2017).

9.4.3 Semiconductor nanoparticles

Other kinds of NPs include semiconductor nanoparticles that have emerged as contrast agents for cell labeling, PA imaging, and in vivo studies of reactive oxygen species characteristic of tumors. Quantum dots are semiconductive nanocrystals that exhibit multifunctionality and are suited for molecular imaging and photothermal therapy due to wide absorption bands, large surface areas, and narrow emission bands (Mallidi, 2011; Onoshima et al., 2015; Yu et al., 2012).

9.4.4 Organic nanoparticles

NPs are modified to be used as a contrast agent for PA imaging. One such study focuses on the development of a 100-nm diameter nanoparticle-containing indocyanine green (ICG). It is based on photonic explorers that use PEBBLE technology and organically modified silicate (ormosil) as a matrix. ICG-embedded ormosil PEBBLEs were highly stable and conjugated with Her-2 antibody for targeting breast cancer and prostate cancer. The successful results indicated that it could be used for detection and photodynamic therapy (Kim et al., 2007).

9.4.5 Inorganic magnetic nanoparticles

Superparamagnetic iron oxide nanoparticle (SPION), a clinically approved metal oxide nanoparticle, has a unique property of superparamagnetism through which they can generate heat in magnetic fields or be guided to a target tissue when an external magnetic field is applied. Thus it is widely used in MRI, targeted tumor annihilation through hyperthermia, tissue engineering, etc. (Singh et al., 2010). The different stages of metastasis have been discussed in earlier sections. As the dissemination occurs in blood during metastasis, the circulating tumor cells are shed into the vasculature and may localize into potential metastatic sites. Recent understanding of

CTCs implicates that they can be used as a marker to predict disease progression and survival in metastatic patients. In addition, CTCs, if a source of metastatic cells, could indicate tumor staging and therapeutic management (Plaks et al., 2013). However, in the case of CTCs, the limited sample volume and low numbers of CTCs pose new challenges for early diagnosis. Hence CTC detection and multiplex targeting strategy has been made possible by conjugating hybrid AuNPs with magnetic NPs to yield better specificity of detection (Galanzha et al., 2009).

9.4.6 Graphene nanomaterial

Graphene-based nanomaterials (GBNs) have unique physicochemical properties and functionalization that have vast applications in biomedicine, diagnostics, and therapy. A combination of chemotherapy and photo-thermal treatment was developed with doxorubicin (DOX)–loaded PEGylated nanographene oxide (NGO–PEG [polyethylene glycol]–DOX) that could potentially deliver heat and drug molecule to the tumor (Zhang et al., 2011). Nanosheets via physical absorption or chemical conjugation have been used to deliver antitumor drugs like doxorubicin, camptothecin, paclitaxel, 1,3-bis(2-chloroethyl)-1-nitrosourea, fluorouracil, methotrexate, lucanthone, b-lapachone, and ellagic acid (Shim et al., 2016). GBNs have gained interest in cancer therapy along with photothermal therapy and photodynamic therapy. The optical properties of GBNs in the visible and NIR range along with their small size, affordability, and low toxicity are attractive and useful targets for bioimaging and photothermal cancer diagnostics and therapy (Shareena et al., 2018).

9.5 Biomedical applications of PAI in cancer

Different imaging modalities like MRI, computer-assisted tomography (CAT), single-photon emission computed tomography, and positron emission tomography have been used in clinics. Nevertheless, the health hazards associated with radiation, high expenses, and inadequate specificity limit their usage in early cancer detection, unlike PAI, which has achieved significant success in cancer diagnostics and staging (Fig. 9.5) (Zou et al., 2017).

PAI was used to image the microvasculature corresponding to angiogenesis in prostate cancer through a novel PAI system equipped with transrectal ultrasound (TRUS) type probe. The PA signal intensity was correlated with microvascular density (MVD), total vascular area (TVA), and total vascular length (TVL) assessed by CD34 immunostaining (Horiguchi et al., 2017). Biocompatible AuNPs loaded with paclitaxel (PTX) were used for drug delivery and PAI of MDA-MB-231 cells, a highly metastatic triple-negative breast cancer cell line. PTX-COS AuNPs exhibited sustained drug release and induced apoptosis of tumor cells that also proved to be a new class of contrast agents for PAI (Manivasagan et al., 2016).

A new photoacoustic mammography (PAMMG) system with a hemispheric detector array was developed to study tumor vasculature in breast cancer patients.

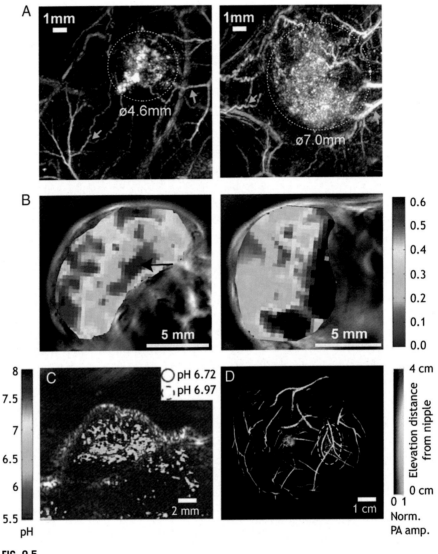

FIG. 9.5

Examples of photoacoustic images of vasculature https://dmm.biologists.org/content/12/7/dmm039636 (Brown et al., 2019).

This system offered superior contrast enabling the visualization of fine vasculature in comparison with standard contrast-enhanced MRI. The overlaid MR-PA images gave a detailed understanding of tumor vasculature and characterization of tumor microenvironment by visualization of oxygen saturation status of hemoglobin at different wavelengths (Toi et al., 2017). PAI has wider applications in breast cancer as the tissue contains fat and is enriched with blood supply or vasculature. The assessment of tumor margins using endogenous contrast agents like fat and hemoglobin in breast cancers is a novel approach. The PA signals generated by tumor regions and normal tissues were differentiated, and tumor margin was determined with histologic analysis. This system offers a submillimeter axial resolution and deep penetration coupled with high imaging speed (Li et al., 2015).

Tumor metastasis occurs on sentinel lymph nodes (SLNs) that receive lymphatic drainage from cancerous tissues. The lack of endogenous contrast molecules like Hb makes it inevitable to use exogenous contrast agents. One such report is of AuNPs endocytosed by mesenchymal stem cells and targeting them to metastatic specific biomarkers. The resultant PA images could identify lymph node metastasis with high sensitivity and specificity (Nam et al., 2012; Pan et al., 2012; Zou et al., 2017)

A combination of therapeutic and diagnostic capabilities using nanoparticles has led to advancements in personalized medicine (Kelkar & Reineke, 2011). The advancements in PA imaging has led to the use of exogenous contrast agents in the diagnosis of pathophysiological process at the cellular and molecular levels using nanomaterials through the generation of high-resolution images for imaging-guided diagnosis and therapy. Despite the success and wide applications of nanoparticle-based probes in PA imaging exhibiting enhanced signals, the challenges posed by the biosafety of their use in a clinical setting is still not well established and validated. An interesting study has investigated ICG, an FDA-approved NIR fluorescence dye-labeled to panitumumab, an FDA-approved monoclonal antibody against human EGFR as a target-based PA imaging probe that was able to enhance target-specific PA signals and study the metastatic potential of tumors (Sano et al., 2015).

Over the past two decades, PA imaging systems have seen many advancements in breast cancer detection. However, there is a need to perform significant number of in vivo studies and clinical studies comprising a larger breast cancer patient pool with different types of breast cancer and staging. With the standardization of PAI-based diagnostic protocols for different cancer signatures and developments in image reconstruction technologies, PA imaging can be a very promising tool in breast cancer screening, diagnosis, and therapy in the future.

References

Akter, H., Park, M., Kwon, O.S., Song, E.J., Park, W.S., Kang, M.J., 2015. Activation of matrix metalloproteinase-9 (MMP-9) by neurotensin promotes cell invasion and migration through ERK pathway in gastric cancer. Tumor Biology 36 (8), 6053–6062.

Andisheh-Tadbir, A., Hamzavi, M., Rezvani, G., Ashraf, M.J., Fattahi, M.J., Khademi, B., Kamali, F., 2014. Tissue expression, serum and salivary levels of vascular endothelial growth factor in patients with HNSCC. Brazilian Journal of Otorhinolaryngology 80 (6), 503–507.

Anselmo, A.C., Mitragotri, S., 2016. Nanoparticles in the clinic. Bioengineering & Translational Medicine 1 (1), 10–29.

Apte, R.S., Chen, D.S., Ferrara, N., 2019. VEGF in signaling and disease: Beyond discovery and development. Cell 176 (6), 1248–1264.

Autio, K.A., Dreicer, R., Anderson, J., Garcia, J.A., Alva, A., Hart, L.L., Milowsky, M.I., Posadas, E.M., Ryan, C.J., Graf, R.P., Dittamore, R., 2018. Safety and efficacy of BIND-014, a docetaxel nanoparticle targeting prostate-specific membrane antigen for patients with metastatic castration-resistant prostate cancer: A phase 2 clinical trial. JAMA Oncology 4 (10), 1344–1351.

Aydın, E.A., Torun, A.R., 2020. 3D printed PLA/copper bowtie antenna for biomedical imaging applications. Physical and Engineering Sciences in Medicine, 1–11.

Bao, C., Conde, J., Curtin, J., Artzi, N., Tian, F., Cui, D., 2015. Bioresponsive antisense DNA gold nanobeacons as a hybrid in vivo theranostics platform for the inhibition of cancer cells and metastasis. Scientific Reports 5, 12297.

Baptista, P.V., 2014. Gold nanobeacons: A potential nanotheranostics platform. Nanomedicine 9 (15), 2247–2250.

Beard, P., 2011. Biomedical photoacoustic imaging. Interface Focus 1 (4), 602–631.

Boudreau, N.J., Jones, P.L., 1999. Extracellular matrix and integrin signalling: The shape of things to come. Biochemical Journal 339 (3), 481–488.

Bray, F., Ferlay, J., Soerjomataram, I., Siegel, R.L., Torre, L.A., Jemal, A., 2018. Global cancer statistics 2018: GLOBOCAN estimates of incidence and mortality worldwide for 36 cancers in 185 countries. CA: A Cancer Journal for Clinicians 68 (6), 394–424.

Brown, E., Brunker, J., Bohndiek, S.E., 2019. Photoacoustic imaging as a tool to probe the tumour microenvironment. Disease Models & Mechanisms 12 (7), dmm039636.

Butler, T.P., Gullino, P.M., 1975. Quantitation of cell shedding into efferent blood of mammary adenocarcinoma. Cancer Research 35 (3), 512–516.

Carioli, G., Malvezzi, M., Rodriguez, T., Bertuccio, P., Negri, E., La Vecchia, C., 2018. Trends and predictions to 2020 in breast cancer mortality: Americas and Australasia. The Breast 37, 163–169.

Chambers, A.F., Groom, A.C., MacDonald, I.C., 2002. Dissemination and growth of cancer cells in metastatic sites. Nature Reviews Cancer 2 (8), 563–572.

Chan, K.T., Cortesio, C.L., Huttenlocher, A., 2009. FAK alters invadopodia and focal adhesion composition and dynamics to regulate breast cancer invasion. Journal of Cell Biology 185 (2), 357–370.

Costache, M.I., Ioana, M., Iordache, S., Ene, D., Costache, C.A., Săftoiu, A., 2015. VEGF expression in pancreatic cancer and other malignancies: A review of the literature. Romanian Journal of Internal Medicine 53 (3), 199–208.

Cubasch, H., Dickens, C., Joffe, M., Duarte, R., Murugan, N., Chih, M.T., Moodley, K., Sharma, V., Ayeni, O., Jacobson, J.S., Neugut, A.I., 2018. Breast cancer survival in Soweto, Johannesburg, South Africa: A receptor-defined cohort of women diagnosed from 2009 to 11. Cancer Epidemiology 52, 120–127.

Dang, Y.Z., Zhang, Y., Li, J.P., Hu, J., Li, W.W., Li, P., Wei, L.C., Shi, M., 2017. High VEGFR1/2 expression levels are predictors of poor survival in patients with cervical cancer. Medicine (1), 96.

De La Zerda, A., Zavaleta, C., Keren, S., Vaithilingam, S., Bodapati, S., Liu, Z., Levi, J., Smith, B.R., Ma, T.J., Oralkan, O., Cheng, Z., 2008. Carbon nanotubes as photoacoustic molecular imaging agents in living mice. Nature Nanotechnology 3 (9), 557–562.

Dewhirst, M.W., Ashcraft, K.A., 2016. Implications of increase in vascular permeability in tumors by VEGF: A commentary on the pioneering work of Harold Dvorak. Cancer Research 76 (11), 3118–3120.

Dhillon, A.S., Hagan, S., Rath, O., Kolch, W., 2007. MAP kinase signalling pathways in cancer. Oncogene 26 (22), 3279–3290.

Ding, F., Zhan, Y., Lu, X., Sun, Y., 2018. Recent advances in near-infrared II fluorophores for multifunctional biomedical imaging. Chemical Science 9 (19), 4370–4380.

Duncan, R., 2003. The dawning era of polymer therapeutics. Nature Reviews Drug Discovery 2 (5), 347–360.

Ewertz, M., Land, L.H., Dalton, S.O., Cronin-Fenton, D., Jensen, M.B., 2018. Influence of specific comorbidities on survival after early-stage breast cancer. Acta Oncologica 57 (1), 129–134.

Fagiani, E., Christofori, G., 2013. Angiopoietins in angiogenesis. Cancer Letters 328 (1), 18–26.

Fang, J., Nakamura, H., Maeda, H., 2011. The EPR effect: unique features of tumor blood vessels for drug delivery, factors involved, and limitations and augmentation of the effect. Advanced Drug Delivery Reviews 63 (3), 136–151.

Ferrara, N., Adamis, A.P., 2016. Ten years of anti-vascular endothelial growth factor therapy. Nature Reviews Drug Discovery 15 (6), 385–403.

Fiedler, U., Augustin, H.G., 2006. Angiopoietins: A link between angiogenesis and inflammation. Trends in Immunology 27 (12), 552–558.

Folkman, J., 1995. Clinical applications of research on angiogenesis. New England Journal of Medicine 333 (26), 1757–1763.

Fukumura, D., Kloepper, J., Amoozgar, Z., Duda, D.G., Jain, R.K., 2018. Enhancing cancer immunotherapy using antiangiogenics: Opportunities and challenges. Nature Reviews Clinical Oncology 15 (5), 325.

Galanzha, E.I., Shashkov, E.V., Kelly, T., Kim, J.W., Yang, L., Zharov, V.P., 2009. In vivo magnetic enrichment and multiplex photoacoustic detection of circulating tumour cells. Nature Nanotechnology 4 (12), 855–860.

Gao, X., Balan, V., Tai, G., Raz, A., 2014. Galectin-3 induces cell migration via a calcium-sensitive MAPK/ERK1/2 pathway. Oncotarget 5 (8), 2077.

Geiger, T.R., Peeper, D.S., 2009. Metastasis mechanisms. Biochimica et Biophysica Acta (BBA)-Reviews on Cancer 1796 (2), 293–308.

Gómez-Cuadrado, L., Tracey, N., Ma, R., Qian, B., Brunton, V.G., 2017. Mouse models of metastasis: Progress and prospects. Disease Models & Mechanisms 10 (9), 1061–1074.

Gonda, A., Zhao, N., Shah, J.V., Calvelli, H.R., Kantamneni, H., Francis, N.L., Ganapathy, V., 2019. Engineering tumor-targeting nanoparticles as vehicles for precision nanomedicine. MedOne, 4.

Graham, M.T., Bell, M.A.L., 2020. Photoacoustic spatial coherence theory and applications to coherence-based image contrast and resolution. IEEE Transactions on Ultrasonics, Ferroelectrics, and Frequency Control 67 (10), 2069–2084.

Gupta, G.P., Massagué, J., 2006. Cancer metastasis: building a framework. Cell 127 (4), 679–695.

Gustafson, H.H., Holt-Casper, D., Grainger, D.W., Ghandehari, H., 2015. Nanoparticle uptake: The phagocyte problem. Nano Today 10 (4), 487–510.

Gwak, J.M., Kim, H.J., Kim, E.J., Chung, Y.R., Yun, S., Seo, A.N., Lee, H.J., Park, S.Y., 2014. MicroRNA-9 is associated with epithelial-mesenchymal transition, breast cancer stem cell phenotype, and tumor progression in breast cancer. Breast Cancer Research and Treatment 147 (1), 39–49.

Harney, A.S., Karagiannis, G.S., Pignatelli, J., Smith, B.D., Kadioglu, E., Wise, S.C., Hood, M.M., Kaufman, M.D., Leary, C.B., Lu, W.P., Al-Ani, G., 2017. The selective Tie2 inhibitor rebastinib blocks recruitment and function of Tie2Hi macrophages in breast cancer and pancreatic neuroendocrine tumors. Molecular Cancer Therapeutics 16 (11), 2486–2501.

Haroon, Z.A., Amin, K., Saito, W., Wilson, W., Greenberg, C.S., Dewhirst, M.W., 2002. SU5416 delays wound healing through inhibition of TGF-β activation. Cancer Biology & Therapy 1 (2), 121–126.

Heer, E., Harper, A., Escandor, N., Sung, H., McCormack, V., Fidler-Benaoudia, M.M., 2020. Global burden and trends in premenopausal and postmenopausal breast cancer: A population-based study. The Lancet Global Health 8 (8), e1027–e1037.

Heijblom, M., Klaase, J.M., Van Den Engh, F.M., Van Leeuwen, T.G., Steenbergen, W., Manohar, S., 2011. Imaging tumor vascularization for detection and diagnosis of breast cancer. Technology in Cancer Research & Treatment 10 (6), 607–623.

Horiguchi, A., Shinchi, M., Nakamura, A., Wada, T., Ito, K., Asano, T., Shinmoto, H., Tsuda, H., Ishihara, M., 2017. Pilot study of prostate cancer angiogenesis imaging using a photoacoustic imaging system. Urology 108, 212–219.

Hu, M., Chen, J., Li, Z.Y., Au, L., Hartland, G.V., Li, X., Marquez, M., Xia, Y., 2006. Gold nanostructures: Engineering their plasmonic properties for biomedical applications. Chemical Society Reviews 35 (11), 1084–1094.

Jacob, A., Prekeris, R., 2015. The regulation of MMP targeting to invadopodia during cancer metastasis. Frontiers in Cell and Developmental Biology 3, 4.

Jiang, H., Shao, W., Zhao, W., 2014. VEGF-C in non-small cell lung cancer: meta-analysis. Clinica Chimica Acta 427, 94–99.

Jiang, Y., Pu, K., 2017. Advanced photoacoustic imaging applications of near-infrared absorbing organic nanoparticles. Small 13 (30), 1700710.

Joneson, T., McDonough, M., Bar-Sagi, D., Van Aelst, L., 1996. RAC regulation of actin polymerization and proliferation by a pathway distinct from Jun kinase. Science 274 (5291), 1374–1376.

Kelkar, S.S., Reineke, T.M., 2011. Theranostics: Combining imaging and therapy. Bioconjugate Chemistry 22 (10), 1879–1903.

Kim, C., Erpelding, T.N., Jankovic, L., Pashley, M.D., Wang, L.V., 2010. Deeply penetrating in vivo photoacoustic imaging using a clinical ultrasound array system. Biomedical Optics Express 1 (1), 278–284.

Kim, G., Huang, S.W., Day, K.C., O'Donnell, M., Agayan, R.R., Day, M.A., Kopelman, R., Ashkenazi, S., 2007. Indocyanine-green-embedded PEBBLEs as a contrast agent for photoacoustic imaging. Journal of Biomedical Optics 12 (4), 044020.

Kim, S.A., Heinze, K.G., Waxham, M.N., Schwille, P., 2004. Intracellular calmodulin availability accessed with two-photon cross-correlation. Proceedings of the National Academy of Sciences 101 (1), 105–110.

Korpanty, G., Sullivan, L.A., Smyth, E., Carney, D.N., Brekken, R.A., 2010. Molecular and clinical aspects of targeting the VEGF pathway in tumors. Journal of Oncology 2010.

Kreibig, U., Vollmer, M., 2013. Optical properties of metal clusters (vol. 25). Springer Science & Business Media.

Kulkarni, S.A., Feng, S.S., 2013. Effects of particle size and surface modification on cellular uptake and biodistribution of polymeric nanoparticles for drug delivery. Pharmaceutical Research 30 (10), 2512–2522.

Lalwani, G., Cai, X., Nie, L., Wang, L.V., Sitharaman, B., 2013. Graphene-based contrast agents for photoacoustic and thermoacoustic tomography. Photoacoustics 1 (3-4), 62–67.

Lashkari, B., Mandelis, A., 2010. Photoacoustic radar imaging signal-to-noise ratio, contrast, and resolution enhancement using nonlinear chirp modulation. Optics Letters 35 (10), 1623–1625.

Lee, K.A., Nelson, C.M., 2012. New insights into the regulation of epithelial-mesenchymal transition and tissue fibrosis. International Review of Cell and Molecular Biology 294, 171–221.

Li, M.L., Wang, J.C., Schwartz, J.A., Gill-Sharp, K.L., Stoica, G., Wang, L.V., 2009. In-vivo photoacoustic microscopy of nanoshell extravasation from solid tumor vasculature. Journal of Biomedical Optics 14 (1), 010507.

Li, R., Wang, P., Lan, L., Lloyd, F.P., Goergen, C.J., Chen, S., Cheng, J.X., 2015. Assessing breast tumor margin by multispectral photoacoustic tomography. Biomedical Optics Express 6 (4), 1273–1281.

Lou, W., Liu, J., Gao, Y., Zhong, G., Chen, D., Shen, J., Bao, C., Xu, L., Pan, J., Cheng, J., Ding, B., 2017. MicroRNAs in cancer metastasis and angiogenesis. Oncotarget 8 (70), 115787.

Mallidi, S., Luke, G.P., Emelianov, S., 2011. Photoacoustic imaging in cancer detection, diagnosis, and treatment guidance. Trends in Biotechnology 29 (5), 213–221.

Mani, S.A., Guo, W., Liao, M.J., Eaton, E.N., Ayyanan, A., Zhou, A.Y., Brooks, M., Reinhard, F., Zhang, C.C., Shipitsin, M., Campbell, L.L., 2008. The epithelial-mesenchymal transition generates cells with properties of stem cells. Cell 133 (4), 704–715.

Manivasagan, P., Bharathiraja, S., Bui, N.Q., Lim, I.G., Oh, J., 2016. Paclitaxel-loaded chitosan oligosaccharide-stabilized gold nanoparticles as novel agents for drug delivery and photoacoustic imaging of cancer cells. International Journal of Pharmaceutics 511 (1), 367–379.

Mathur, P., Sathishkumar, K., Chaturvedi, M., Das, P., Sudarshan, K.L., Santhappan, S., Nallasamy, V., John, A., Narasimhan, S., Roselind, F.S.ICMR-NCDIR-NCRP Investigator Group, 2020. Cancer statistics, 2020: Report from National Cancer Registry Programme, India. JCO Global Oncology 6, 1063–1075.

McKinney, S.M., Sieniek, M., Godbole, V., Godwin, J., Antropova, N., Ashrafian, H., Back, T., Chesus, M., Corrado, G.C., Darzi, A., Etemadi, M., 2020. International evaluation of an AI system for breast cancer screening. Nature 577 (7788), 89–94.

Miao, L., Huang, L., 2015. Exploring the tumor microenvironment with nanoparticlesNanotechnology-based precision tools for the detection and treatment of cancer. Springer, Cham, pp. 193–226.

Minchinton, A.I., Tannock, I.F., 2006. Drug penetration in solid tumours. Nature Reviews Cancer 6 (8), 583–592.

Mittendorf, E.A., Bartlett, J.M., Lichtensztajn, D.L., Chandarlapaty, S., 2018. Incorporating biology into breast cancer staging: American Joint Committee on Cancer, revisions and beyond. American Society of Clinical Oncology Educational Book 38, 38–46.

Nam, S.Y., Ricles, L.M., Suggs, L.J., Emelianov, S.Y., 2012. In vivo ultrasound and photoacoustic monitoring of mesenchymal stem cells labeled with gold nanotracers. PLoS One 7 (5), e37267.

Nishida, N., Yano, H., Nishida, T., Kamura, T., Kojiro, M., 2006. Angiogenesis in cancer. Vascular Health and Risk Management 2 (3), 213.

O'Neal, D.P., Hirsch, L.R., Halas, N.J., Payne, J.D., West, J.L., 2004. Photo-thermal tumor ablation in mice using near infrared-absorbing nanoparticles. Cancer Letters 209 (2), 171–176.

Onoshima, D., Yukawa, H., Baba, Y., 2015. Multifunctional quantum dots-based cancer diagnostics and stem cell therapeutics for regenerative medicine. Advanced Drug Delivery Reviews 95, 2–14.

Pan, D., Cai, X., Yalaz, C., Senpan, A., Omanakuttan, K., Wickline, S.A., Wang, L.V., Lanza, G.M., 2012. Photoacoustic sentinel lymph node imaging with self-assembled copper neodecanoate nanoparticles. ACS Nano 6 (2), 1260–1267.

Pan, D., Pramanik, M., Senpan, A., Allen, J.S., Zhang, H., Wickline, S.A., Wang, L.V., Lanza, G.M., 2011. Molecular photoacoustic imaging of angiogenesis with integrin-targeted gold nanobeacons. The FASEB Journal 25 (3), 875–882.

Pan, D., Pramanik, M., Senpan, A., Yang, X., Song, K.H., Scott, M.J., Zhang, H., Gaffney, P.J., Wickline, S.A., Wang, L.V., Lanza, G.M., 2009. Molecular photoacoustic tomography with colloidal nanobeacons. Angewandte Chemie 121 (23), 4234–4237.

Panigrahy, D., Kaipainen, A., Butterfield, C.E., Chaponis, D.M., Laforme, A.M., Folkman, J., Kieran, M.W., 2010. Inhibition of tumor angiogenesis by oraletoposide. Experimental and Therapeutic Medicine 1 (5), 739–746.

Pardal, R., Clarke, M.F., Morrison, S.J., 2003. Applying the principles of stem-cell biology to cancer. Nature Reviews Cancer 3 (12), 895–902.

Plaks, V., Koopman, C.D., Werb, Z., 2013. Circulating tumor cells. Science 341 (6151), 1186–1188.

Pouliot, N., Pearson, H.B., Burrows, A., 2013. Investigating metastasis using in vitro platformsMadame Curie Bioscience Database [Internet]. Landes Bioscience.

Risau, W., 1997. Mechanisms of angiogenesis. Nature 386 (6626), 671–674.

Roger, E., Lagarce, F., Garcion, E., Benoit, J.P., 2010. Biopharmaceutical parameters to consider in order to alter the fate of nanocarriers after oral delivery. Nanomedicine 5 (2), 287–306.

Rostro-Kohanloo, B.C., Bickford, L.R., Payne, C.M., Day, E.S., Anderson, L.J., Zhong, M., Lee, S., Mayer, K.M., Zal, T., Adam, L., Dinney, C.P., 2009. The stabilization and targeting of surfactant-synthesized gold nanorods. Nanotechnology 20 (43), 434005.

Sanna, V., Pala, N., Sechi, M., 2014. Targeted therapy using nanotechnology: focus on cancer. International Journal of Nanomedicine 9, 467.

Sano, K., Ohashi, M., Kanazaki, K., Ding, N., Deguchi, J., Kanada, Y., Ono, M., Saji, H., 2015. In vivo photoacoustic imaging of cancer using indocyanine green-labeled monoclonal antibody targeting the epidermal growth factor receptor. Biochemical and Biophysical Research Communications 464 (3), 820–825.

Sarrió, D., Rodriguez-Pinilla, S.M., Hardisson, D., Cano, A., Moreno-Bueno, G., Palacios, J., 2008. Epithelial-mesenchymal transition in breast cancer relates to the basal-like phenotype. Cancer Research 68 (4), 989–997.

Senger, D.R., Van De Water, L., Brown, L.F., Nagy, J.A., Yeo, K.T., Yeo, T.K., Berse, B., Jackman, R.W., Dvorak, A.M., Dvorak, H.F., 1993. Vascular permeability factor (VPF, VEGF) in tumor biology. Cancer and Metastasis Reviews 12 (3-4), 303–324.

Shareena, T.P.D., McShan, D., Dasmahapatra, A.K., Tchounwou, P.B., 2018. A review on graphene-based nanomaterials in biomedical applications and risks in environment and health. Nano-Micro Letters 10 (3), 53.

Shibuya, M., 2011. Vascular endothelial growth factor (VEGF) and its receptor (VEGFR) signaling in angiogenesis: a crucial target for anti-and pro-angiogenic therapies. Genes & Cancer 2 (12), 1097–1105.

Shim, W.S., Ho, I.A., Wong, P.E., 2007. Angiopoietin: A TIE (d) balance in tumor angiogenesis. Molecular Cancer Research 5 (7), 655–665.

Shim, G., Kim, M.G., Park, J.Y., Oh, Y.K., 2016. Graphene-based nanosheets for delivery of chemotherapeutics and biological drugs. Advanced Drug Delivery Reviews 105, 205–227.

Siegel, R.L., Miller, K.D., Jemal, A., 2020. Cancer statistics, 2020. CA: A Cancer Journal for Clinicians 70 (1), 7–30.

Singh, N., Jenkins, G.J., Asadi, R., Doak, S.H., 2010. Potential toxicity of superparamagnetic iron oxide nanoparticles (SPION). Nano Reviews 1 (1), 5358.

Singh, H., Milner, C.S., Hernandez, A., Patel, N., Brindle, N.P.J., 2009. Vascular endothelial growth factor activates the Tie family of receptortyrosine kinases. Cellular Signalling 21 (8), 1346–1350.

Sinn, H.P., Kreipe, H., 2013. A brief overview of the WHO classification of breast tumors. Breast Care 8 (2), 149–154.

Skrabalak, S.E., Au, L., Li, X., Xia, Y., 2007. Facile synthesis of Ag nanocubes and Au nanocages. Nature Protocols 2 (9), 2182–2190.

Son, J., Yi, G., Yoo, J., Park, C., Koo, H., Choi, H.S., 2019. Light-responsive nanomedicine for biophotonic imaging and targeted therapy. Advanced Drug Delivery Reviews 138, 133–147.

Song, K., Wang, L.V., 2007. Deep reflection-mode photoacoustic imaging of biological tissue. Journal of Biomedical Optics 12 (6), 060503.

Srabovic, N., Mujagic, Z., Mujanovic-Mustedanagic, J., Softic, A., Muminovic, Z., Rifatbegovic, A., Begic, L., 2013. Vascular endothelial growth factor receptor-1 expression in breast cancer and its correlation to vascular endothelial growth factor a. International Journal of Breast Cancer *2013*.

Sun, Y., Xia, Y., 2002. Shape-controlled synthesis of gold and silver nanoparticles. Science 298 (5601), 2176–2179.

Tiede, S., Meyer-Schaller, N., Kalathur, R.K.R., Ivanek, R., Fagiani, E., Schmassmann, P., Stillhard, P., Häfliger, S., Kraut, N., Schweifer, N., Waizenegger, I.C., 2018. The FAK inhibitor BI 853520 exerts anti-tumor effects in breast cancer. Oncogenesis 7 (9), 1–19.

Todorova, D., Simoncini, S., Lacroix, R., Sabatier, F., Dignat-George, F., 2017. Extracellular vesicles in angiogenesis. Circulation Research 120 (10), 1658–1673.

Toi, M., Asao, Y., Matsumoto, Y., Sekiguchi, H., Yoshikawa, A., Takada, M., Kataoka, M., Endo, T., Kawaguchi-Sakita, N., Kawashima, M., Fakhrejahani, E., 2017. Visualization of tumor-related blood vessels in human breast by photoacoustic imaging system with a hemispherical detector array. Scientific Reports 7, 41970.

Valastyan, S., Weinberg, R.A., 2011. Tumor metastasis: Molecular insights and evolving paradigms. Cell 147 (2), 275–292.

van Hellemond, I.E., Geurts, S.M., Tjan-Heijnen, V.C., 2018. Current status of extended adjuvant endocrine therapy in early stage breast cancer. Current Treatment Options in Oncology 19 (5), 26.

van Zijl, F., Krupitza, G., Mikulits, W., 2011. Initial steps of metastasis: Cell invasion and endothelial transmigration. Mutation Research/Reviews in Mutation Research 728 (1-2), 23–34.

Vicent, M.J., Ringsdorf, H., Duncan, R., 2009. Polymer therapeutics: Clinical applications and challenges for development. Advanced Drug Delivery Reviews 61 (13), 1117.

Virani, N.A., Davis, C., McKernan, P., Hauser, P., Hurst, R.E., Slaton, J., Silvy, R.P., Resasco, D.E., Harrison, R.G., 2017. Phosphatidylserine targeted single-walled carbon nanotubes for photothermal ablation of bladder cancer. Nanotechnology 29 (3), 035101.

Weber, J., Beard, P.C., Bohndiek, S.E., 2016. Contrast agents for molecular photoacoustic imaging. Nature Methods 13 (8), 639–650.

Wright, J.A., Richer, J.K., Goodall, G.J., 2010. microRNAs and EMT in mammary cells and breast cancer. Journal of Mammary Gland Biology and Neoplasia 15 (2), 213–223.

Wu, Y., Zhou, B.P., 2008. New insights of epithelial-mesenchymal transition incancer metastasis. Acta Biochimica et Biophysica Sinica 40 (7), 643–650.

Xie, H., Pallero, M.A., Gupta, K., Chang, P., Ware, M.F., Witke, W., Kwiatkowski, D.J., Lauffenburger, D.A., Murphy-Ullrich, J.E., Wells, A., 1998. EGF receptor regulation of cell motility: EGF induces disassembly of focal adhesions independently of the motility-associated PLC gamma signaling pathway. Journal of Cell Science 111 (5), 615–624.

Yang, Z., Tian, R., Wu, J., Fan, Q., Yung, B.C., Niu, G., Jacobson, O., Wang, Z., Liu, G., Yu, G., Huang, W., 2017. Impact of semiconducting perylene diimide nanoparticle size on lymph node mapping and cancer imaging. ACS Nano 11 (4), 4247–4255.

Yi, M., Jiao, D., Qin, S., Chu, Q., Wu, K., Li, A., 2019. Synergistic effect of immune checkpoint blockade and anti-angiogenesis in cancer treatment. Molecular Cancer 18 (1), 60.

Yi, M., Jiao, D., Xu, H., Liu, Q., Zhao, W., Han, X., Wu, K., 2018. Biomarkers for predicting efficacy of PD-1/PD-L1 inhibitors. Molecular Cancer 17 (1), 1–14.

Yoo, S.Y., Kwon, S.M., 2013. Angiogenesis and its therapeutic opportunities. Mediators of Inflammation 11, Article ID 127170.

Yu, J., Wu, C., Zhang, X., Ye, F., Gallina, M.E., Rong, Y., Wu, I.C., Sun, W., Chan, Y.H., Chiu, D.T., 2012. Stable functionalization of small semiconducting polymer dots via covalent cross-linking and their application for specific cellular Imaging. Advanced Materials 24 (26), 3498–3504.

Zang, J., Li, C., Zhao, L.N., Shi, M., Zhou, Y.C., Wang, J.H., Li, X., 2013. Prognostic value of vascular endothelial growth factor in patients with head and neck cancer: A meta-analysis. Head & Neck 35 (10), 1507–1514.

Zhang, F., Li, C., Liu, H., Wang, Y., Chen, Y., Wu, X., 2014. The functional proteomics analysis of VEGF-treated human epithelial ovarian cancer cells. Tumor Biology 35 (12), 12379–12387.

Zhang, Y., Hong, H., Cai, W., 2011. Photoacoustic imaging. Cold Spring Harbor Protocols 2011 (9), pdb–top065508.

Zhao, J., Wallace, M., Melancon, M.P., 2014. Cancer theranostics with gold nanoshells. Nanomedicine 9 (13), 2041–2057.

Zou, C., Wu, B., Dong, Y., Song, Z., Zhao, Y., Ni, X., Yang, Y., Liu, Z., 2017. Biomedical photoacoustics: Fundamentals, instrumentation and perspectives on nanomedicine. International Journal of Nanomedicine 12, 179.

Zuazo-Gaztelu, I., Casanovas, O., 2018. Unraveling the role of angiogenesis in cancer ecosystems. Frontiers in Oncology 8, 248.

https://dmm.biologists.org/content/12/7/dmm039636.

https://www.researchgate.net/figure/VEGF-signaling-interactions-The-VEGF-family-can-bind-to-VEGFR1-VEGFR2-and-VEGFR3_fig1_45185768.

https://dmm.biologists.org/content/10/9/1061.

http://www.scientificanimations.com/wiki-images/.

https://dmm.biologists.org/content/12/7/dmm039636.

Further reading

Attia, A.B.E., Balasundaram, G., Moothanchery, M., Dinish, U.S., Bi, R., Ntziachristos, V., Olivo, M., 2019. A review of clinical photoacoustic imaging: Current and future trends. Photoacoustics 16, 100144.

Balasundaram, G., Ho, C.J.H., Li, K., Driessen, W., Dinish, U.S., Wong, C.L., Ntziachristos, V., Liu, B., Olivo, M., 2015. Molecular photoacoustic imaging of breast cancer using an actively targeted conjugated polymer. International Journal of Nanomedicine 10, 387.

Cao, F., Qiu, Z., Li, H., Lai, P., 2017. Photoacoustic imaging in oxygen detection. Applied Sciences 7 (12), 1262.

Choi, W., Park, E.Y., Jeon, S., Kim, C., 2018. Clinical photoacoustic imaging platforms. Biomedical Engineering Letters 8 (2), 139–155.

Cox, B.T., Laufer, J.G., Beard, P.C., Arridge, S.R., 2012. Quantitative spectroscopic photoacoustic imaging: A review. Journal of Biomedical Optics 17 (6), 061202.

Cox, B.T., Laufer, J.G., Beard, P.C., 2009. The challenges for quantitative photoacoustic imagingPhotons plus ultrasound: Imaging and sensing 20097177. International Society for Optics and Photonics, 717713.

Fan, Y., Mandelis, A., Spirou, G., Alex Vitkin, I., 2004. Development of a laser photothermoacoustic frequency-swept system for subsurface imaging: Theory and experiment. The Journal of the Acoustical Society of America 116 (6), 3523–3533.

Fu, Q., Zhu, R., Song, J., Yang, H., Chen, X., 2019. Photoacoustic imaging: contrast agents and their biomedical applications. Advanced Materials 31 (6), 1805875.

Haroon, Z.A., Hettasch, J.M., Lai, T.S., Dewhirst, M.W., Greenberg, C.S., 1999. Tissue transglutaminase is expressed, active, and directly involved in rat dermal wound healing and angiogenesis. The FASEB Journal 13 (13), 1787–1795.

Hoelen, C.G.A., De Mul, F.F.M., Pongers, R., Dekker, A., 1998. Three-dimensional photoacoustic imaging of blood vessels in tissue. Optics Letters 23 (8), 648–650.

Hu, Y., Wang, R., Wang, S., Ding, L., Li, J., Luo, Y., Wang, X., Shen, M., Shi, X., 2016. Multifunctional Fe_3O_4@ Au core/shell nanostars: A unique platform for multimode imaging and photothermal therapy of tumors. Scientific Reports 6 (1), 1–12.

Jain, R.K., 2014. Antiangiogenesis strategies revisited: from starving tumors to alleviating hypoxia. Cancer Cell 26 (5), 605–622.

Jathoul, A.P., Laufer, J., Ogunlade, O., Treeby, B., Cox, B., Zhang, E., Johnson, P., Pizzey, A.R., Philip, B., Marafioti, T., Lythgoe, M.F., 2015. Deep in vivo photoacoustic imaging of mammalian tissues using a tyrosinase-based genetic reporter. Nature Photonics 9 (4), 239–246.

Jokerst, J.V., Cole, A.J., Van de Sompel, D., Gambhir, S.S., 2012. Gold nanorods for ovarian cancer detection with photoacoustic imaging and resection guidance via Raman imaging in living mice. ACS Nano 6 (11), 10366–10377.

Jordan, V.C., 2020. Molecular mechanism for breast cancer incidence in the Women's Health Initiative. Cancer Prevention Research 13 (10), 807–816.

Karsy, M., Guan, J., Sivakumar, W., Neil, J.A., Schmidt, M.H., Mahan, M.A., 2015. The genetic basis of intradural spinal tumors and its impact on clinical treatment. Neurosurgical Focus 39 (2), E3.

Kim, S., Chen, Y.S., Luke, G.P., Emelianov, S.Y., 2011. In vivo three-dimensional spectroscopic photoacoustic imaging for monitoring nanoparticle delivery. Biomedical Optics Express 2 (9), 2540–2550.

Kolkman, R.G., Steenbergen, W., van Leeuwen, T.G., 2006. In vivo photoacoustic imaging of blood vessels with a pulsed laser diode. Lasers in Medical Science 21 (3), 134–139.

Kong, X., Wu, S.H., Zhang, L., Chen, X.Q., 2017. Pilot application of lipoxin A4 analog and lipoxin A4 receptor agonist in asthmatic children with acute episodes. Experimental and Therapeutic Medicine 14 (3), 2284–2290.

Koskinen, J.P., Färkkilä, N., Sintonen, H., Saarto, T., Taari, K., Roine, R.P., 2019. The association of financial difficulties and out-of-pocket payments with health-related quality of life among breast, prostate and colorectal cancer patients. Acta Oncologica 58 (7), 1062–1068.

Lao, Y., Xing, D., Yang, S., Xiang, L., 2008. Noninvasive photoacoustic imaging of the developing vasculature during early tumor growth. Physics in Medicine & Biology 53 (15), 4203.

Laufer, J.G., Zhang, E.Z., Treeby, B.E., Cox, B.T., Beard, P.C., Johnson, P., Pedley, B., 2012. In vivo preclinical photoacoustic imaging of tumor vasculature development and therapy. Journal of Biomedical Optics 17 (5), 056016.

Lee, J.C., 2008. Modulation of allostery of pyruvate kinase by shifting of an ensemble of microstates. Acta Biochimica et Biophysica Sinica 40 (7), 663–669.

Levi, J., Kothapalli, S.R., Bohndiek, S., Yoon, J.K., Dragulescu-Andrasi, A., Nielsen, C., Tisma, A., Bodapati, S., Gowrishankar, G., Yan, X., Chan, C., 2013. Molecular photoacoustic imaging of follicular thyroid carcinoma. Clinical Cancer Research 19 (6), 1494–1502.

Liu, Y., Bhattarai, P., Dai, Z., Chen, X., 2019. Photothermal therapy and photoacoustic imaging via nanotheranostics in fighting cancer. Chemical Society Reviews 48 (7), 2053–2108.

Liu, Y., Nie, L., Chen, X., 2016. Photoacoustic molecular imaging: from multiscale biomedical applications towards early-stage theranostics. Trends in Biotechnology 34 (5), 420–433.

Li, W., Chen, X., 2015. Gold nanoparticles for photoacoustic imaging. Nanomedicine 10 (2), 299–320.

Luke, G.P., Yeager, D., Emelianov, S.Y., 2012. Biomedical applications of photoacoustic imaging with exogenous contrast agents. Annals of Biomedical Engineering 40 (2), 422–437.

Maneas, E., Aughwane, R., Huynh, N., Xia, W., Ansari, R., Singh, Kuniyil Ajith, M., Hutchinson, J., C., Sebire, N.J., Arthurs, O.J., Deprest, J., Ourselin, S, 2020. Photoacoustic imaging of the human placental vasculature. Journal of Biophotonics 13 (4), e201900167.

Manohar, S., Gambhir, S.S., 2020. Clinical photoacoustic imaging. Photoacoustics, 19.

Moore, C., Jokerst, J.V., 2019. Strategies for image-guided therapy, surgery, and drug delivery using photoacoustic imaging. Theranostics 9 (6), 1550.

Olafsson, R., Bauer, D.R., Montilla, L.G., Witte, R.S., 2010. Real-time, contrast enhanced photoacoustic imaging of cancer in a mouse window chamber. Optics Express 18 (18), 18625–18632.

O'Shaughnessy, J., Brezden-Masley, C., Cazzaniga, M., Dalvi, T., Walker, G., Bennett, J., Ohsumi, S., 2020. Prevalence of germline BRCA mutations in HER2-negative metastatic breast cancer: global results from the real-world, observational BREAKOUT study. Breast Cancer Research 22 (1), 1–11.

Palicharla, V.R., Maddika, S., 2015. HACE1 mediated K27 ubiquitin linkage leads to YB-1 protein secretion. Cellular Signalling 27 (12), 2355–2362.

Pericleous, P., Gazouli, M., Lyberopoulou, A., Rizos, S., Nikiteas, N., Efstathopoulos, E.P., 2012. Quantum dots hold promise for early cancer imaging and detection. International Journal of Cancer 131 (3), 519–528.

Piras, D., Xia, W., Steenbergen, W., van Leeuwen, T.G., Manohar, S., 2009. Photoacoustic imaging of the breast using the twente photoacoustic mammoscope: Present status and future perspectives. IEEE Journal of Selected Topics in Quantum Electronics 16 (4), 730–739.

Qu, B., Zhang, X., Han, Y., Peng, X., Sun, J., Zhang, R., 2020. IR820 functionalized melanin nanoplates for dual-modal imaging and photothermal tumor eradication. Nanoscale Advances.

Senger, D.R., Van De, W.L., 1993. Vascular endothelial growth factor in tumor biology. Cancer Metastasis Review 12 (6), 303–311.

Siphanto, R.I., Thumma, K.K., Kolkman, R.G.M., Van Leeuwen, T.G., De Mul, F.F.M., Van Neck, J.W., Van Adrichem, L.N.A., Steenbergen, W., 2005. Serial noninvasive photoacoustic imaging of neovascularization in tumor angiogenesis. Optics Express 13 (1), 89–95.

Su, J.L., Wang, B., Wilson, K.E., Bayer, C.L., Chen, Y.S., Kim, S., Homan, K.A., Emelianov, S.Y., 2010. Advances in clinical and biomedical applications of photoacoustic imaging. Expert Opinion on Medical Diagnostics 4 (6), 497–510.

Valluru, K.S., Wilson, K.E., Willmann, J.K., 2016. Photoacoustic imaging in oncology: Translational preclinical and early clinical experience. Radiology 280 (2), 332–349.

Wang, S., Lin, J., Wang, T., Chen, X., Huang, P., 2016. Recent advances in photoacoustic imaging for deep-tissue biomedical applications. Theranostics 6 (13), 2394.

Wang, Y., Xing, D., Zeng, Y., Chen, Q., 2004. Photoacoustic imaging with deconvolution algorithm. Physics in Medicine & Biology 49 (14), 3117.

Webb, A.G., 2017. Introduction to biomedical imaging. John Wiley & Sons.

Wilson, K.E., Wang, T.Y., Willmann, J.K., 2013. Acoustic and photoacoustic molecular imaging of cancer. Journal of Nuclear Medicine 54 (11), 1851–1854.

Xi, L., Grobmyer, S.R., Wu, L., Chen, R., Zhou, G., Gutwein, L.G., Sun, J., Liao, W., Zhou, Q., Xie, H., Jiang, H., 2012. Evaluation of breast tumor margins in vivo with intraoperative photoacoustic imaging. Optics Express 20 (8), 8726–8731.

Xu, M., Wang, L.V., 2006. Photoacoustic imaging in biomedicine. Review of Scientific Instruments 77 (4), 041101.

Yang, X., Stein, E.W., Ashkenazi, S., Wang, L.V., 2009. Nanoparticles for photoacoustic imaging. Wiley Interdisciplinary Reviews: Nanomedicine and Nanobiotechnology 1 (4), 360–368.

Yang, Z., Song, J., Tang, W., Fan, W., Dai, Y., Shen, Z., Lin, L., Cheng, S., Liu, Y., Niu, G., Rong, P., 2019. Stimuli-responsive nanotheranostics for real-time monitoring drug release by photoacoustic imaging. Theranostics 9 (2), 526.

Zerda, A.D.L., Liu, Z., Bodapati, S., Teed, R., Vaithilingam, S., Khuri-Yakub, B.T., Chen, X., Dai, H., Gambhir, S.S., 2010. Ultrahigh sensitivity carbon nanotube agents for photoacoustic molecular imaging in living mice. Nano Letters 10 (6), 2168–2172.

Near-infrared spectroscopy: An important noninvasive and sensitive tool for point-of-care biosensing application

10

Subhavna Juneja[a], Ranjita Ghosh Moulick[b], Deepak Kushwaha[b], Harsh A Gandhi[a], Jaydeep Bhattacharya[a]

[a]*School of Biotechnology, Jawaharlal Nehru University, New Delhi*
[b]*Amity Institute of Integrative Science and Health/ Amity Institute of Biotechnology, Amity University, Gurgaon, Haryana*

10.1 Introduction

Near-infrared spectroscopy (NIS) refers to the analytic technique that deals with the electromagnetic spectrum in the near-infrared (NIR) region (780–2500 nm). It characterizes the molecule of interest by studying its interaction with light falling in this spectrum. This region of light, identified with peculiar properties, interacts readily with different bond types such as −NH, −CH, and −OH, characterized with specific frequencies or wavelengths (Beć & Huck, 2019). As NIR falls on the sample, depending upon the chemical nature of the compound, energies corresponding to constituting bonds are absorbed, giving information about the analyte under study. NIS is universal to analysis for liquid, semisolid, and most solid samples.

Typically, NIS is a vibrational spectroscopy subtype yet it differs from its other forms such as far IR, mid IR, and Raman spectroscopy where the information is built on fundamental transitions only. Internuclear potential energies and atomic bond modeling have been described by drawing similarities to a harmonic oscillator, which when displaced from equilibrium experiences a restoring force proportional to its displacement (Pasquini, 2003). Although this analogy works to a certain extent for low energy transitions, it fails to model all atomic bonds, dissociations, and additional vibrational frequencies observed in the vibrational spectra simply because the bonds do not behave like a harmonic oscillator in reality but are instead found to be associated with additional vibrations that fall away from the fundamental ones. NIS is the study of reading these additional weaker spectral lines arising due to bond anharmonicity, giving it an analytic distinctiveness. The resonant restoring frequencies lying above the fundamental frequencies are referred to as molecular overtones, and are sample specific. IRS characters can thus be understood as multiples of the fundamental absorbing frequency (Hills, 2017).

Biomedical Imaging Instrumentation. DOI: https://doi.org/10.1016/B978-0-323-85650-8.00004-8

Unlike other vibrational analytic techniques that allow harmonic approximation to comprehend and relate to spectral information theoretically, NIS built on anharmonic approach had limited computational assistance, thus limiting its utilization in a practical world. Nonetheless, electronic and information technology (IT) advancements in the present scenario have made it possible to program intensive anharmonicity of complex samples, making it feasible to extend the study beyond some basic model analytes, explaining emerging trends in NIR analysis. NIS has distinguished itself from a mere vibrational spectral technique to sought as one of the most important analytic tools having contributions to different fields, including medicine, physical chemistry, material science, food technology, agricultural sciences, astronomy, and anharmonicity modeling, etc., as discussed in detail in the following segments of the chapter (Chen et al., 2020; Cortés et al., 2019; Grassi & Alamprese, 2018; Ozaki, 2012; Sakudo, 2016).

10.2 Fundamentals of near-infrared spectroscopy

Each frequency of light has a certain energy, where infrared covers the range of the electromagnetic spectrum between 1 and 100 μm, which corresponds to the energy range between 8 and 40 kJ/mole (Harris, 2007; Hershel, 1800; Hollas, 2004; Van Mass, 1972).

10.2.1 Basic divisions in infrared region

IR region can be mainly divided into following three basic zones depending on the spectrum as shown in Fig. 10.1.

However, depending on sensitivity of the detector, IR region is classified into five zones instead of the three according to sensor response division scheme (Hesse et al., 1997; Pavia & Lampman, 2001; Rijeka, 2012). Some classifications also include regions of thermal IR (TIR: 8–15 μm) or long-wave IR (LWIR).

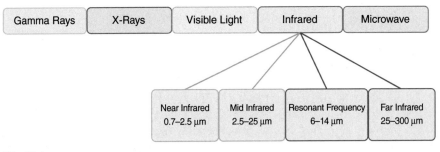

FIG. 10.1

The electromagnetic spectrum showing regions of IR.

10.2.2 **Molecular overtones and near-infrared spectra**

Molecular overtones and combinations are the two major NIR absorption processes. Fundamental frequency and its overtones are known as harmonic partials. An overtone is a result of molecular excitation to the second excited state that generates a series of integer multiples of the fundamental frequency (Roggo et al., 2007). A higher energy compared to fundamental and few overtones is observed in case of NIR absorptions. In contrast combinations appear due to allocation of energies by excitation between two or more fundamental mid-IR (MIR) bands. Hence, the use of NIR is limited due to the region of broad peaks. To resolve the complexity of the spectra, multivariate calibration techniques are used (Martens & Naes, 1996).

10.3 **Instrumentation of near-infrared spectrophotometer**

10.3.1 **Instrumentation**

The fundamental design of an NIR instrument is in principle similar to an optical absorption spectrophotometer. Hence the scheme of design is essentially the one used in popular forms such as visible (Vis) or IR spectroscopy. In fact, certain instruments specified for Vis or IR spectroscopy can measure at least some part in the NIR region. Such measurements depend on emission spectrum of the source and sensitivity of the detector. Based on these two factors, ultraviolet (UV) Vis spectrometers are often capable of measuring in short-wavelength NIR region. Similarly, most dedicated IR spectrometers are able to measure in the long-wavelength NIR region. Nevertheless, to obtain a full spectrum of greater quality and information, a dedicated NIR instrument is required with a specified light source (e.g., tungsten halogen lamp) and a detector (e.g., high-performing indium gallium arsenide [InGaAs]) (Brown, 2007).

The block design of an NIR spectrometer constitutes a radiation source, a wavelength selector or interferometer and a detector, interfaced by optical components (Fig. 10.2A). Based on type of optics, NIR spectrometers can be categorized into two different types: wavelength dispersive and Fourier transform (FT). The original dispersive type IR spectroscopy instruments perform IR dispersion by means of a prism or a grating. Similar to a prism used to split visible light, an infrared prism breaks down an IR light into its constituent frequencies. A grating is slightly advanced material, which works better in splitting of an IR ray. The energy of each frequency that has passed through the sample is measured by a detector and presented as a frequency versus intensity plot (Armstrong et al., 2006; Tran, 2005). A notable drawback of IR dispersion instruments is slow scanning speed due to simultaneous detection and measurement of frequencies. The solution to this problem came with the development of an elegant optical device called an interferometer, which was able to produce signal encoding all the possible IR frequencies. Most spectrometers use a Michelson interferometer, which constitutes an optical unit formed by a fixed and a moving mirror (see Fig. 10.2B).

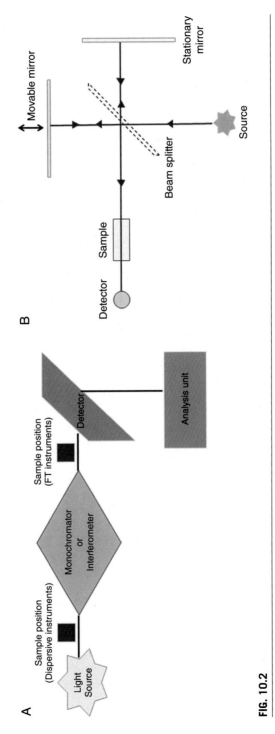

FIG. 10.2

(A) Basic components in IR spectrophotometer assembly, (B) schematics of a Michelson-type interferometer.

Using a beam splitter, a polychromatic beam is simultaneously refracted toward both the mirrors. The reflected lights from the mirrors meet at the beam splitter, resulting in a recombined wave, which is directed toward the sample. The path difference between the beams introduced over the time of the scan leads to periodically alternating interferences (phase differences). This gives an interferogram from which a spectrum can be reconstructed using Fourier transformation. The signal from the interferometer can be measured nearly in the time scale of a second. This reduces the time component per sample from several minutes to a few seconds (Perkins, 1986).

As mentioned previously, the main components of an NIR spectrometer are a radiation source, a sample optical system, a spectrometer, and a detector. Broadly, two types of NIR instruments are classified based on the arrangement of system components mentioned earlier. In one type, the sample is placed after the spectrometer and irradiated with monochromatic light. In the other type, the sample is positioned just after the light source and irradiated with white light. The type of system components may vary from instrument to instrument. A particular NIR instrument may have one of various possible combinations of the components depending on the system requirements. Table 10.1 enlists different component types used in NIR instruments. For detailed classification of instrument based on wavelength selection technology, please refer to Pasquini (2003, 2018).

Table 10.1 Different types of components in NIR spectroscopy.

S.No.	Near Infrared Light Sources	Spectroscopic Elements	Detectors	Other Optical Materials	Ref.
1.	Thermal Radiation (Halogen Tungsten Lamps)	PRISM (Dispersion)	Silicone Photo Diode (Si)	Transparent Material for NIR Light	(Pasquini, 2003, 2018)
2.	Light Emitting Diodes	Diffraction Grating (Dispersion)	Indium Gallium Arsenide (InGaAs)	High- and Low- Reflection Materials	
3.	LASER Diodes	Fourier Transform	Lead Sulfide (PbS)	Polarizer	
4.	Solid State LASERS	Bandpass Filters -Interference Filter -Variable Filters -Liquid Crystal Tunable Filters (LCTF) -Acousto-optic Tunable Filters (AOTF)	Photomultiplier Tube (PMT)	— —	
5.	Supercontinuum Light	— —	Image Sensor (Multichannel Detector)	— —	

10.3.2 Sampling techniques and measuring methods in near-infrared spectroscopy

NIR instrument may use one of the following four methods to capture the absorption spectrum (Bendoula & Ducanchez, 2019; Gastélum-Barrios et al., 2020).

a. **Transmission or transmittance method:** In this method, light is passed through a cuvette or NIR transparent material, and the transmitted light is captured into the input slit. This method is generally used for NIR transparent substances such as plastics, films, fuels, oils, polymers, liquids, solvents, and solutions. The usage of light by these systems is most efficient because light is immediately captured by the input slit after transmission through the sample.

b. **Reflection or reflectance method:** This method is used for materials that are not transparent to NIR such as solids, powders, and polymers. Light is directed onto the surface of these materials, and reflected light is captured into the input slit. However, this is not as simple as it seems because in many cases some light from the incident beam is reflected by the surface and some of it is transmitted beneath the surface. The transmitted light is refracted and reflected many times on the surface of the particles (milled grains, suspensions, cells, etc.), and a part of it goes outside the scatterer. This is called the diffuse reflection (DR) light and is comparable to the transmittance light because the DR ray travels inside the absorptive material. Due to the dispersive nature of the material in reflective method, more light power is required than the transmissive method.

c. **Transflectance method:** This method is basically a combination of transmission and reflection methods. Light is reflected by a background material after passing through the sample. The reflected light is passed again through the sample, hence passing twice through the sample. Transflectance method is typically used for gels, liquids, polymers, slurries, and pastes.

d. **Fiber optic interfacing:** In case of remote sampling, fiber optic cable bundles are used. In a typical setup, a fiber bundle is divided into two sets, one of which is directed to a light source. The second set captures the reflected light and sends it back into the input slit.

10.3.2.1 Contact and noncontact-based sampling

In contact-based sampling, the light and collection optics are in close contact resulting in minimal loss of signal. Since the reflected light originates from the inside of the material, the internal composition is majorly elucidated and not the surface one.

Some NIR applications, however, require the contactless sampling and measurement to avoid the sample contamination between and across the samples. In such noncontact sampling, due to separation between sample and light optics, less light is provided to the sample. Hence these systems require added light/power to minimize the background noise. Both surface and internal composition of the analyte can be characterized by this method (Wold et al., 2006).

10.3.2.2 *Sample preparation for near-infrared analysis*

To minimize data degeneracy, sample homogeneity is mandatory. Each sample type may have specific requirements depending upon its physical state. For example, the liquids may be sampled as a uniformly dispersed solution, while the solids may be crushed, ground, pulverized, and mixed to obtain a solid pellet.

In particular cases where the thickness of sample may affect the scan, solids are cut or sliced to a specific thickness. Gases must be sampled in a gas cell with sufficiently longer pathlength to enable the detection of the compound. Temperature and pressure might need to be well controlled so as to maintain the density of a gaseous sample.

10.3.2.3 *Selection of a reference*

A reference is an external standard that is used to compare the absorbance with transmittance or reflectance. Different references might be needed for different types of samples (Agelet & Hurburgh, 2010).

a. **References for transparent samples:** The references for transparent samples should be insensitive to NIR radiation. Any material free of scattering centers such as a solid transparent film or a clear solution can be referenced as blank. However, most solvents show strong absorption in the NIR region and often overlap with the absorption of the target solute, making it necessary to note the solvent type during quantitative analysis.

b. **References for opaque samples:** Most solids show diffuse reflection (DR); hence it is difficult to choose right reference material for such materials. For a good quality NIR scan it is highly desirable to remove the scattering component. To this end, the scattering coefficient of the reference should be similar to that of the sample, and the reference should not show absorbance in the objective NIR region. Polytetrafluorethylene (PTFE), ceramic plate, and gold are such commonly used reference materials.

10.4 **Analyzing an infrared spectrum**

The use of interferometer in FT instruments provides an edge over the dispersive instruments. At the time of its advent, FT technology required high-end component electronics and supercomputers for data analysis. The expensive machinery and high maintenance thus limited the use of this technology. However, after the microelectronic revolution, a significant improvement and price reduction increased the availability of FT instruments. FT IR offers several advantages such as better scan speed, improved signal-to-noise ratio, reproducibility, accuracy, and high resolution. The major influencing attributes of FT IR are its **multiplexing** (where all major frequencies can be continually measured to create a multifaceted spectrum) and **throughput** (whereby use of an interferometer increased the output by providing a large orifice) (Banwell & McCash, 1994; Landau, 1976; Prati et al., 2016; Schrader, 1995; Skoog et al., 2007)

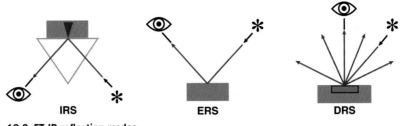

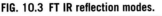

FIG. 10.3 FT IR reflection modes.

IRS (Internal Reflectance Spectroscopy); ERS (External Reflectance Spectroscopy); DRS (Diffuse Reflectance Spectroscopy).

(Adapted from Marten & Naes, 1996).

Measurement of samples by FT or dispersive instruments broadly depends on transmittance and reflectance properties of the sample. Transmission techniques often require slight preprocessing of samples; for example, solids might need to be mixed with potassium bromide (KBr) and packed together in a compressed disc. For liquid samples, special holders mounted on a surface of a KBr disc are used. In the reflectance method, samples do not generally require special pretreatments, and data are recorded based on reflective properties. However, depending on the composition of the sample, reflection-based measurement technique can be classified in three major types (Fig. 10.3):

a. **Attenuated total reflectance IR (ATR/IRS):** Several internal reflections occur in an ATR crystal having high refractive index (RI) and generate an evanescent wave that spreads to the sample. The wave infiltrates to a particular depth depending on several controls such as RI of the crystal and wavelength of light used. Though ATR method is compatible with a wide variety of samples, it is limited to use in NIR owing to weak overtone and combination band vibration (Banwell & McCash, 1994; Landau, 1976).

b. **Specular reflection spectroscopy (SRS/ERS):** It is used when light is reflected from a specular surface (mirrorlike) at an angle equal to the angle of incidence. Reflectance spectra is analogous to ATR IR thin films' spectra and identified as a reflection absorption mode. When an IR spectrometer is coupled with an optical microscope to collect the specular reflection, it is called a micro-FT IR (Massart et al., 2003; Skoog et al., 2007).

c. **Diffuse reflectance spectroscopy (DRS):** Light, when it interacts with rough surfaces such as powders, undergoes both external and internal reflections. The resultant reflections in all directions form a diffuse spectrum that gives information about the analyte. An example of such method is DRIFT (diffuse reflectance infrared Fourier transform), where a diffuse reflectance is used with FT IR spectroscopy (Banwell & McCash, 1994; Perkin Elmer Life and Analytical Sciences, 2005; Prati et al., 2016).

10.4.1 Data treatment of an infrared spectrum

The plot of absorption intensity versus the wave number or wavelength gives IR spectra of the compound. It is represented as peaks of all the relative absorptions.

Position of a band depends on the mass of atoms and bond strength, whereas the strength and width of a band depends on change in dipole moment and hydrogen bonding, respectively.

Unlike MIR spectrum, which is made up of peak intensities and positions, chemometric NIR data consists of absorption reflections from overtones and fundamental vibration combinations. Multivariate analysis is an ideal approach for studying the NIR spectrum due to its ability of handling multicomponents in less time. There are three most common approaches used to examine the NIR data (Banwell & McCash, 1994; Landau, 1976; Massart, 2003; Perkin Elmer Life and Analytical Sciences, 2005; Prati et al., 2016; Schrader, 1995; Skoog et al., 2007):

a. **Arithmetic data preprocessing:** This approach uses techniques of derivatization, normalization, and smoothing. It keeps the focus on major peaks in the spectrum that helps in reducing the impact of auxiliary or minor peaks.

b. **Classification techniques:** This technique is used for the qualitative data analysis. A supervised classification is applied where prior knowledge of samples' categories is available, which can be used to create a classification model. This model is then evaluated by comparing actual and predicted values. Partial least square discriminant analysis (PLSDA) is a classic supervised approach. In unsupervised methods where no other information is available except for the spectra, principal component analysis (PCA) can be used (Banwell & McCash, 1994; Landau, 1976; Perkin Elmer Life and Analytical Sciences, 2005; Prati et al., 2016; Schrader, 1995; Skoog et al., 2007).

c. **Regression methods:** These are mainly used for quantitative analysis. Principal component regression (PCR), artificial neural network (ANN), support vector machines (SVMs), and multilinear regression are common approaches (Roggo et al., 2007).

10.5 General applications of near-infrared spectroscopy

NIR spectroscopy is not particularly a sensitive technique, yet it has found implementation and acceptance based on little to no requirement of sample preparation supported by improved optics, informatics, and new scientific development such as nanostructures. NIR as a fingerprint assay finds use in the following fields.

Agriculture and environmental science: Assessing crop variety, crop health, soil quality, disease identification, and protein expression.

Food science: Detection of food adulteration, identification of transgenic food, meat processing, fish quality, rind assessment in grated Parmigiano-Reggiano cheese, moisture content, and sensory (texture, color, flavor) features.

Pharmaceutic industry: Identification and detection of different drugs and small molecules (active/excipients), validation studies under quality control, moisture content estimation, identification of drug homogeneity, and determination of optical polymorphs.

Clinical chemistry: Assessment of disease-associated biochemical changes, deviations from homeostasis, and physiologic characterization of body fluids provides information on biologic molecules such as nucleic acids, carbohydrates, proteins and peptides, lipids, and others. For example, the measurement of cholesterol and triglycerides in blood plays a major role in managing the modern lifestyle disease. NIR spectroscopy is capable of noninvasive measurement of this kind of small biomolecules directly from body fluids.

10.5.1 **Non invasive technology in clinical chemistry**

a. Glucose monitoring: Most diabetic patients use blood glucose self-monitoring (SMBG) devices to keep their glucose levels in check and to adjust their insulin dosage to achieve normoglycemia. Needle-type wearable sensors based on enzyme-mediated electrochemistry are now available for continuous monitoring of interstitial glucose within the subcutaneous skin tissue, but they suffer from systematic deviations from the capillary or arterial blood glucose level as gold standard for treatment with insulin. Compared to SMBG, a noninvasive measurement system addresses the invasiveness of traditional needle-type sensors allowing much higher reading frequency and elimination of physical pain (Karoui et al., 2010).

b. Skin optical data–mediated photon migration modeling: The diffuse radiation transport in biologic tissues can be modeled by using different mathematical tools providing an understanding, for example, of the probed tissue volume and photon penetration depths. The mean optical path length for radiation within mucosa tissue, as given for diffuse reflectance accessories, is wave number dependent. Tissue optical properties are the absorption and the scattering coefficients, μa and μs (in units of mm^{-1}), respectively, and the anisotropy of scattering g (dimensionless). From the latter two parameters, the reduced scattering coefficient $\mu s' = \mu s (1 - g)$ can be calculated. From diffusion theory for photon transport in tissues, an optical penetration depth can be estimated based on these optical constants with $\delta = [3\,\mu a\,(\mu a + \mu s')]^{-1/2}$ (Delbeck et al., 2019).

c. Hemoglobin and blood ethanol monitoring: The importance of hemoglobin (Hb) measurements has already been highlighted when presenting in vitro assays. However, there are scenarios, where it is desirable to monitor hemoglobin continuously as during transfusion, for patients under intensive care or with a postoperative follow-up. In principle, the time-resolved signals from photoplethysmography are evaluated with the observed alternating current (AC) and direct current (DC) signals measured. The selected radiation is found within the visible and short-wave (SW) NIR spectral range with wavelengths between 600 and 1000 nm, realized, for example, by a light-emitting diode (LED) array with center wavelengths of 569, 660, 805, 940, and 975 nm (selected wavelengths represent two isosbestic points and three for compensation of tissue scattering). A more sophisticated model considers also the scattering of the arterial blood arising from the red blood cells (Kessoku et al., 2011). Further improvements were achieved by designing a special finger probe with optimizations of the detector area, the emission area of a light source, and the distance between the light source and the detector. Such an optimally designed finger probe provided a correlation coefficient of 0.869 and a standard deviation of 0.81 g/dL in predicting total Hb (Olesberg et al., 2006).

d. **Blood tissue oxygenation and cytochrome redox studies:** Besides the determination of blood glucose, there are further special biomolecules of great interest for monitoring deviations from homeostasis in the body. Natural pigments such as Hb as the major component in the red blood cells, myoglobin in muscle, as well as cytochromes generally found in tissues, have characteristic absorbance spectra within the visible and the SW NIR spectral range (Delbeck et al., 2019). Their spectra are dependent on the degree of oxygenation of the hemoproteins and the redox state of the cytochromes. Therefore spectroscopy can provide information about the in-vivo state of tissue oxygen supply.

NIR Imaging: Emission in NIR facilitates biologic imaging based on its ability to penetrate deeper in the tissue and minimal interference from molecular milieu, improving sensitivity. It finds application in detection of biologically relevant species such as enzymes, proteins, mRNAs, etc., both in vitro and in-vivo. NIR has been used to screen reactive oxygen/nitrogen species, sensing metal ions, thiols, hydrogen sulfide, and intracellular pH changes.

10.6 Nanomaterial-assisted near-infrared spectroscopy–based biosensing

Among different classes of contrasting agents available, using NIR responsive fluorophores for bioanalysis can significantly improve signals as NIR penetrates deeper than UV or visible light. It is nonionizing and minimally invasive (Manley, 2014). They permit improved imaging depth and higher spatial resolution when compared to conventional linear optical approach. Moreover, autofluorescence and background interference are minimal as water and major absorbing tissues such as blood are transparent in the 700 to 1100 nm spectral window. NIR phosphor categories extending into organic molecules, dyes, and other small molecules supported developing the modality as a powerful diagnostic tool that it is today; however, development of nonlinear optics and utilization of nanomaterial-based NIR fluorophores completely revolutionized the field. Tunable optical properties, high specific surface areas, photochemical stability, resistance to photobleaching, high quantum yields, sharp emission features, and large Stokes shift (Liu et al, 2016) of the nanoparticles all make them a preferred fluorophore system. Additionally, their multiphoton excitation, multiplexing sensing ability, and combinational therapeutics add weight to their versatility and performance abilities. Nanoparticle-based fluorophores are available in abundance, both in number and types, since no one system satiates all the criterion necessary for it to be an ideal fluorescing system. Having said so, there might be no one fit for all nanoparticle systems, but each type might be ideal for a specific application under study. This section of the chapter describes four major classes of nanomaterial fluorophores, their key attributes, uses, and limitations toward biosensing/imaging application. The mentioned nanomaterial fluorophores are the most common, but they are not limited to these categories. Detailing all the different nanoparticle fluorescing systems is beyond the technical limits of the chapter, and we wish to refer readers to articles/reviews in Table 10.2 for further reading.

Table 10.2 List of suggested articles for deep reading into different nanomaterial types used in NIR-based biosensing.

S. No	Nanomaterial type	Advantages	Disadvantages	References
1.	Semiconductor quantum dots	High quantum yield, no photobleaching, band gap–dependent optical properties	Heavy metal toxicity, prolonged body clearance time	(Martynenko et al., 2017; Massey et al., 2015; Zhao P et al., 2018)
2.	Carbon-based nanomaterials	Cellular benignity, indigenous fluorescence, size-dependent electrical/optical tunable properties	Retention in body tissues, tailoring limitations	(Heller et al., 2005; Mazrad, 2018; Pan et al., 2017)
3.	Plasmonic nanostructures	Ease of synthesis, chemical and physical stability, functionalization diversity; size and morphology dependent tunable properties	Prolonged residing duration, thermal heating	(Kim et al., 2017; Muhammed et al., 2015)
4.	Dye encapsulated nanoparticles	Improved brightness, regressed photoquenching, biocompatibility	Free radical release, polymer swelling, resorb ability, loss of payload	(Chapman & Patonay, 2019; Cheng et al., 2020)
5.	Magnetic nanoparticles	Photocatalytic potential, high mass transference, chemical stability	Nanoparticle aggregation, stability	(Urries et al., 2014)
6.	Lanthanide and chalcogenide based	Narrow band absorption/emission spectra, long luminescence decay, multiple labeling, efficient NIR to visible upconversion	Low quantum yield (<3%), surface quenching, intense excitation/emission wavelength optimization	(Fan & Zhang, 2019; Ning et al., 2019; Tan et al., 2018)
7.	Combined reviews	Various	Various	(Altınoğlu & Adair, 2010; Cai Y et al., 2019; Liu et al., 2016; Zhao J et al., 2018)

10.6.1 **Semiconductor quantum dots (QDs)**

Nanoscale clusters of semiconductor materials ideally made up of atoms of the order 10^{2-5} classify as QDs. Their typical size ranges above the atomic scale yet they are small enough to hold charge carrier quantum confinement in all three spatial dimensions (Sargent, 2005). The energy levels are quantized, and their optical/ electrical properties are governed by their size. Quantum dots share an inverse relationship with band gap, where smaller QDs emit in high-energy regions and larger ones emit in lower energy (Alivisatos, 1996; Michalet et al., 2001). They are commonly made up of II to VI (CdS/Se/Te, ZnO/S), III to V (GaAs/N/P, InAs/P), IV to VI (PbS/Se) semiconductors or their alloys (CdSeS/Te, InGaAs).

QDs garnered great scientific and technical inclination for their use in NIR spectroscopy owing to their significant advantages over conventionally used fluoro-phores, which included high quantum yield (>50%), larger extinction coefficients, and brighter stable fluorescence. In a study spread over 2 years, Petryayeva and Algar (2013, 2014) for the first time demonstrated the feasibility of using QDs for a fluo-rescence resonance energy transfer (FRET)–based point-of-care (POC) proteolytic assay. The paper-based substrates were simply readable using everyday gadgets such as smartphones or webcams. The substrates were economical, fast, and accurate, fea-tures especially beneficial for POC application. They also demonstrated multiplex-ing detection ability of their developed paper substrates by simultaneously studying three different proteases (trypsin, chymotrypsin, and enterokinase). Red, blue, and green (RBG) color-emitting QD-protein conjugates were labeled with appropri-ate organic dye acceptors forming FRET systems that were sensitive to proteolytic cleavage capable of detection via the RBG channel intensities of smartphone camera images (Fig. 10.4).

Another recent advancement to the semiconductor QD technology has been the development of core shell dots, where a wider band gap semiconductor encases the nanocrystal, improving their photostability and resistance to bleaching. Resultant fluorescence in the core shell system is a consequence of the band offset of the ma-terials making up the nanomaterial. Such QDs are proven to longer fluorescence lifetimes and energy emission smaller than the actual band gap of any one of the constituting materials (Park et al., 2011). Several health concerns citing the use of heavy metals for QD synthesis, their long stability and residing tendency in the or-ganelles causing cytotoxicity, and apoptosis have raised uncertainty towards their long-term use for in vivo imaging (Valizadeh et al., 2012). Modifications, including oligomeric coatings, nanoshells, to alloyed semiconductors QDs, have been reported (Cai G. et al., 2019; Kim et al., 2004; Zhang et al., 2019). Although, initial studies do not show signs of toxicity, researchers predict that prolonged existence in biological fluids can cause partial disruption of the outer layer exposing cells and tissues to bare QDs triggering adverse immune response (Lalancette-Hébert et al., 2010). Nonethe-less, QDs can find preferable use as in vitro fluorophores where long-term toxicity does not have detrimental outcomes. Superior fluorescence, long-term tracking, and enhanced sensitivity outscore its applicability in non–in vivo setting.

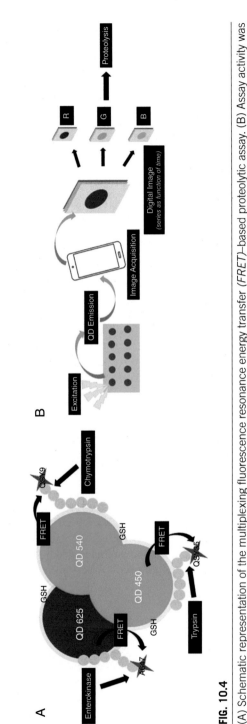

FIG. 10.4

(A) Schematic representation of the multiplexing fluorescence resonance energy transfer *(FRET)*–based proteolytic assay. (B) Assay activity was measured via recovery of quantum dot *(QD)* emission experienced upon dye cleavage and associated increase in signal strength of the dye.

(Adapted from Petryayeva & Algar, 2013, 2014).

10.6.2 Carbon nanomaterials

Outer layer valence four electrons in carbon atoms can form linkages within or with different elements giving it the ability to polymerize at atomic scales. This atomic scale polymerization allows forming long chain structures but with smaller size and peculiar electronic configuration in carbon forms compared to other family compounds in group IV (Zaytseva & Neumann, 2016). Other than diamond and graphite, naturally occurring allotropes of carbon are rare, and most commonly synthesized artificially. Carbon elements thus synthesized, such that one of their dimensions falls between 1 and 100 nm, form, what we refer to as carbon nanoparticles. These can be synthesized using heat activation, vapor deposition, grinding, or chemically through carbonization. They are characterized with $[2s^2\ 2p^2]$ electronic configuration and most commonly are classified based on their geometric features such as nanohorns for horn-shaped carbon nanomaterial, nanotubes for tubular nanostructures, and spheres/ellipsoids (which fall under fullerenes). Their high specific area, intrinsic electrical conductivity, and dynamic atomic polymerization network present their applicability in numerous applications (Cao & Rogers, 2009; Lee et al., 2010; Rahman & Mieno, 2015; Ray & Jana, 2017; Yang et al., 2016).

Nanodiamonds (ND) have been characterized to be inherently rich in luminescence defects with the majority being nitrogen vacancies (Gruber et al., 1997). Ascribed to their nonbleaching 700-nm fluorescence and high quantum yields, they have been utilized for in vivo imaging, cellular diagnosis, and flow cytometry. The substitutional nitrogen vacancy in NDs can also be altered to green or blue fluorescence by altering defect type to N-V-N or manipulating size and functionality expanding their applicability for practical use. Their expanded fluorescence lifetimes, mass production, and benignity to adverse reproductive potentiality and regenerative functions have been recently realized. In a contemporary study by Zhang et al. (2014), they fabricated nanostructured diamond functionalized biosensing electrodes for real-time bioanalyte *(E. coli 0157:H7)* monitoring. The ND integrated electrodes were found to serve as electrically conductive nanoislands with high surface areas and sufficient hydrogenation for antibody binding, assisting increased bacterial cell capture and consequently improved impedance signals. For a cell concentration of 10^6 CFU/mL, resistance values decreased by approximately 38%, which were considerably high compared to traditional redox probe studies.

Numerous research groups have also directed their efforts toward surface engineering clusters of carbon in the form of carbon QDs and nanotubes (CNTs). Excitation-mediated optical properties of QDs make them a successful fluorophore for multicolor imaging applications, while CNTs are characterized with strong NIR II fluorescence in addition to their therapeutic properties. Gogoi and Raju Khan (2018) synthesized an upconversion NIR carbon QD for application in glutathione (GSH) detection. Synthesized carbon QDs were absorptive to 930 nm photons with subsequent emission in 520 nm. For GSH detection, complexation of Cu^{2+} ions with GSH was used to reverse the fluorescence character of the QD. In presence of Cu^{2+} ions, QDs suffered severe quenching in luminescence resulting in hindering the upconversion green emission. However, in the presence of analyte, GSH, Cu^{2+} complexes with GSH reinstating the anti-Stokes emission in the carbon QD. The developed sensor was selective and sensitive at the same time. The limit of detection (LOD) was tested to be as low as 0.35 μM. Wang et al. (2017) synthesized carbon QDs functionalized chitosan nanogels for NIR imaging and pH-sensitive drug release for therapy. The QD-infused nanogels were tested to be highly stable, bright, had high drug loading capacity, and had efficient photothermal properties. Their nontoxicity was proven through various in vitro and in vivo analysis along with their drug release properties that were equally efficient to both the stimuli, NIR and pH. In a study conducted by Kruss et al. (2014), single-walled carbon nanotubes (SWCNTs) were functionalized by suspending in a complex mixture of 30 different polymers, including phospholipids and nucleic acids. The functionalized CNTs were then used to identify neurotransmitters temporally and spatially with neural networks. Carbon-based nanostructures are by far a safer alternative to semiconducting QDs; however, cellular toxicities and tailoring difficulties are major concerns. In fabrication of uniformly dimensional CNTs, specifically, length limits the flexibility of geometric design and thereby optical properties for statistically relevant results.

10.6.3 Plasmonic nanoparticles and nanoshells

Plasmonic nanostructures are small particles, most commonly metallic (Au/Ag/Cu/Pd) in nature, capable of coupling with electromagnetic radiations larger than their size owing to their electron density. These particles act as excellent light scatterers and absorbers (Eustis & El-Sayed, 2006). Similar to QD optical sensitivity to its band gap, plasmonic nanoparticles are optically tunable by altering their shape and size. Their optical response can be tuned from UV to visible to NIR regions of the electromagnetic spectrum.

Plasmonic nanoparticles have found extensive use in energy (De Aberasturi et al., 2015), surface enhanced Raman spectroscopy (Demirel et al., 2012), diagnosis (Sotiriou, 2013), antibacterial coatings (Endo-Kimura & Kowalska, 2020), among other areas. Plasmonic nanostructures, specifically gold, are an attractive alternative as NIR fluorophores due to their benignity, chemical inertness, easy synthesis, optical characteristics, biocompatibility, and tunable properties. A simplistic yet sensitive biosensing platform based on photoelectric reaction was devised by Li et al. (2018) for quantitative determination of cancer cells and cell surface glycans. A gold nanoparticle modified Ag_2S coated ITO slides were used for building the sensor substrate. Water-dispersible Ag_2S QDs served as the photoelectrochemically active species, while gold nanoparticles acted as nanoantennae. On gold coated slides, the conversion efficiency of the electrochemical reaction was found to be increased by 2.5 times compared to bare QDs. Later, 4-mercaptophenylboronic acid (MPBA), working as recognition molecules, were bound to gold nanoparticles through thiol linkages to help detect presence of cancer cells by making use of the sialic acid click chemistry between cytomembranes and MPBA. The experimental limit of detection was tabulated to be 100 cells/mL (Li et al., 2018). In a similar study, Li et al. (2016) fabricated a carcinoembryonic antigen (CEA) sensor using palladium nanoparticles

(PdNPs). The said system was sensitive to CEA at concentration as low as 1.7 pg/mL, observed in diluted human serum sample (Li et al., 2016).

In comparison to the other nanoscale and non-nanoscale NIR fluorophores, the absorption behavior in a metallic plasmonic nanostructure is governed by its morphology rather than electronic transitions (Wang et al., 2004). Thus, as long as the structure is stable, fluorescence is constant. Consequently, plasmonic nanostructures are also less prone to photobleaching and not susceptible to thermal or chemical degradation. Gold nanostructures as such are low on cytotoxicity and resistant to surface oxidation and corrosion; however, its functionalization with polymers such as PEG is known to further suppress immunogenic responses improving circulation time (Harris et al., 2001). Nonetheless, accumulation of even the nontoxic exogenous agents can interfere with traditional biological routine, and thus clearance of plasmonic nanoparticle is an issue that requires immediate attention for its use in in vivo bioimaging. Also, the gold nanoparticles are prone to thermal heating. A sustained temperature rise in the cells can result in triggered immunologic responses and shock heating of proteins. Although this technique is beneficial and has been used for therapy (photothermal), intense local heating during nontherapeutic applications could induce toxicity (Yang et al., 2004).

10.6.4 Dye encapsulating nanoparticles

Dye encapsulated nanoparticles are carriers engineered to cart one or more fluorescent molecules, typically enclosed within an optically clear matrix. The fluorophore-nanoparticle complex works as a single fluorescing feature that assists improved bioimaging/sensing capabilities due to its enhanced photostability. The encapsulate stays protected from any undesirable conformational changes owing to reduced interaction with the surrounding media, also limiting dynamic relaxation pathways and thereby nonradiative energy losses (Avnir et al., 1984). Although encapsulated dye molecules bear phenomenologically similar radiative feature as nonencapsulated dye molecules, the composites are reported to yield higher brightness ($\sim$1–2 orders). Improved brightness is ascribed to improved surface area ratios in nanoparticles that allow multiple dye molecule loading onto a single particle. Overall, it facilitates improved emission features and absorptivity (Muddana et al., 2009). Encapsulation also mitigates photoquenching, a common problem associated with most of the conventional fluorophore systems used in optical imaging.

Sub 100-nm dye encapsulating silica nanoparticles were one of the earliest explored nanoparticle-dye complexes owing to its ease of synthesis, multiple doping routes, photochemical inertness, optical transparency, and well-established silane chemistry-based functionalization for bioapplications (Santra & Dutta, 2007). Moreover, its structural rigidity and porous nature allow efficient caging of the encapsulate overcoming porosity, leakage, or swelling concerns across multiple solvent systems. Lee et al. (2009) synthesized NIR sensitive mesoporous silica nanostructures (MSN), which were modified with trimethylammonium (TA) groups and electrostatically bound to indocyanine dye (ICG) for use as NIR contrast agent. The TA-MSN-ICG complex was observed to be fairly stable over physiologic pH conditions with acceptable biodistribution and fluorescence. Tested in an anesthetized rat model, this formed one of the earliest reports on NIR-ICG-nanoparticle complexes for in vivo optical imaging. Guével et al. (2011) synthesized novel bovine serum albumin–coated ultrasmall gold nanoclusters (<2 nm), which were loaded into porous silica nanoparticles to explore its potential as a biolabel for sensing and imaging of cancerous cells. Metal nanoclusters with high Stokes shift and emission in NIR maintained the fluorescence properties of the composite while silica coating served as a protective coating to improve the photochemical stability of the fluorescent cluster. Despite the offered biocompatibility and surface properties, silica nanoparticles are known for their toxicity and bioclearance issues. Associated with free radical release on the particle surface as well as by the particle digesting phagocytic cells, they are found to initiate renal disorders limiting their usage. Polymer nanoparticles emerged and quickly escalated to be a promising alternative to silica nanoparticles because of their resorbability, biocompatibility, and addressal of issues such as rapid blood clearance and fluorophore degradation (Soppimath et al., 2001).

Perylene dyes, often used as a fluorophore in live cell imaging, is identified with water solubility, nonspecific binding, aggregation caused quenching, and small Stokes shift decreasing its efficacy as a contrasting agent. Pal et al. (2018) doped folic acid conjugated polymer nanoparticles with 2-(2-ethylhexyl)-8-(phenanthren-9-ylethynyl)-1H-benzo[10,5] anthrax[2,1,9-def]-isoquinoline-1,3(2H)-dione (PMIAP) doped folic acid modified polymer nanoparticles to develop NIR emitting live cell probe capable of sensing folate receptor expression in HeLa and MCF-7 cells. Conjugate was characterized by poor aggregation, 230-nm Stokes shift, and high fluorescence and stability. This study paved a route to developing CSR protein targeting for diagnosis.

Jin et al. (2011) synthesized a NIR fluorescent polymer dot by introducing NIR sensitive dye, silicon 2,3-naphthalocyanine bis(trihexylsilyloxide) (NIR775), into the poly(9,9-dioctylfluorene-co-benzothiadiazole) (PFBT) polymer matrix. The polymer dot shared narrowed emission and brighter fluorescence compared to 800-nm emitting quantum dot. Effective encapsulation of NIR 775 into the polymer matrix introduced hydrophobicity reducing the drawbacks of fluorescence quenching due to aggregation. Additionally, light sensitivity of the PFBT polymer facilitated effective energy transfer from polymer to dye dramatically enhancing the brightness of the conjugate complex. The fabricated conjugate presents applicability in in vivo fluorescence imaging and cellular labeling (Fig. 10.5). Au et al. (2013) conjugated thiophene-functionalized oligo-(ethylene glycol)-methacrylate (OEGMA) and sterically stabilized polypyrrole (PPy) nanoparticles to form NIR sensitive contrast agent for optical coherence tomography (OCT).

Several imaging polymer alternatives abound in conventional literature; however, they fall short of widespread applicability in detection for POC devices due to significant loss of payload content during circulation in the body, susceptibility to swelling, and poor matrix shielding and fluorescence intensity (Singh & Mukherjee, 2005). Moreover, their signal stability cannot be controlled and thus is of less use for in vivo imaging. An exceptional alternative capable of addressing the concerns of a polymer

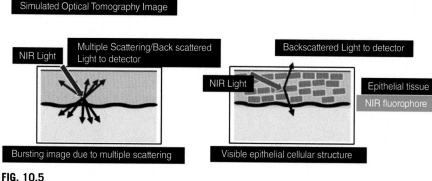

FIG. 10.5

Schematic representation of optical pathway in epithelial tissues in presence and absence of NIR absorbing NIR775—PBFT conjugate nanocomposite.

(Adapted from Jin et al., 2011).

nanoparticle are calcium phosphate nanoparticles (CPN). These nanoparticles are characterized by rigidity of silica, stable fluorescence, insolubility under physiologic pH conditions, and safe biodistribution being a biomineral. Though CPNs are stable at physiologic pH, they are soluble under acidic conditions, which helps in their clearance from the body as a natural process (Tycko & Maxfield, 1982). With emphasis on developing early detection tools, Altınogˇ lu et al. (2008) synthesized bioresorbable CPNs conjugated NIR emitting indocyanine green dye for in vivo breast cancer imaging. The conjugated nanoparticles were exceptionally stable and optically superior. Per molecule quantum efficiency of the conjugated CPNs was 200% higher over free fluorophore. Haedicke et al. (2015) co-functionalized CPNs with a fluorescent dye (DY682-NHS), a photosensitizer (temoporfin), and tumor favoring peptide (RGDfK). The conjugated nanoparticles were injected into tumor bearing CAL-27 model mice. After CPNs–photodynamic therapy (PDT) uptake in mice, considerable decrease in tumor vascularization and volume was detected with triggering of apoptotic processes. Here, CPNs not only aided visualization of the process but also exhibited therapeutic properties by initiating apoptosis and subsequent tumor destruction.

References

Agelet, L.E., Hurburgh, C.R., 2010. A tutorial on near infrared spectroscopy and its calibration. Critical Reviews in Analytical Chemistry 40 (4), 246–260.

Alivisatos, A.P., 1996. Semiconductor clusters, nanocrystals, and quantum dots. Science 271 (5251), 933–937.

Altınoğlu, E.İ., Russin, T.J., Kaiser, J.M., Barth, B.M., Eklund, P.C., Kester, M., Adair, J.H., 2008. Near-infrared emitting fluorophore doped calcium phosphate nanoparticles for in vivo imaging of human breast cancer. ACS Nano 2 (10), 2075–2084.

Altınoğlu, E.İ., Adair, J.H., 2010. Near infrared imaging with nanoparticles. Wiley Interdisciplinary Reviews of Nanomedicine and Nanobiotechnology 2, 461–477.

Armstrong, P.R., Maghirang, E., Xie, F., Dowell, F., 2006. Comparison of dispersive and Fourier-transform NIR instruments for measuring grain and flour attributes. Applied Engineering in Agriculture, 22.

Au, K.M., Lu, Z., Matcher, S.J., Armes, S.P., 2013. Anti-biofouling conducting polymer nanoparticles as a label-free optical contrast agent for high resolution subsurface biomedical imaging. Biomaterials 34-35, 8925–8940.

Avnir, D., Levy, D., Reisfeld, R., 1984. The nature of the silica cage as reflected by spectral changes and enhanced photostability of trapped rhodamine 6G. Journal of Physical Chemistry 88, 5956–5959.

Banwell, C.N., McCash, E.M., 1994. Fundamentals of molecular spectroscopy. McGraw Hill.

Beć, K.B., Huck, C.W., 2019. Breakthrough potential in near-infrared spectroscopy: Spectra simulation. A review of recent developments. Frontiers in Chemistry 7, 48.

Bendoula, R., Ducanchez, A., 2019. Effect of the architecture of fiber-optic probes designed for soluble solid content prediction in intact sugar beet slices. Sensors (13), 19.

Brown, C., 2007. Ultraviolet, visible, and near-infrared spectrophotometers. Applied Spectroscopy Reviews 35 (3), 151–173.

Cai, G., Yu, Z., Tong, P., Tang, D., 2019. Ti_3C_2 MXene quantum dot-encapsulated liposomes for photothermal immunoassays using a portable near-infrared imaging camera on a smartphone. Nanoscale 11, 15659–15667.

Cai, Y., Zheng Wei, Z., Song, C., Tang, C., Han, W., Dong, X., 2019. Optical nano-agents in the second near-infrared window for biomedical applications. Chemical Society Reviews 48, 22–37.

Cao, Q., Rogers, J.A., 2009. Ultrathin films of single-walled carbon nanotubes for electronics and sensors: A review of fundamental and applied aspects. Advanced Materials 21 (1), 29–53.

Chapman, G., Patonay, G., 2019. NIR-fluorescent multidye silica nanoparticles with large Stokes shifts for versatile biosensing applications. Journal of Fluorescence 29, 293–305.

Chen, C., Tian, R., Zeng, Y., Chu, C., Gang Liu, G., 2020. Activatable fluorescence probes for "turn-on" and ratiometric biosensing and bioimaging: From NIR-I to NIR-II. Bioconjugate Chemistry 31 (2), 276–292.

Cheng, Q., Tuanwei Li, T., Tian, Y., Dang, H., Qian, H., Teng, C, Xie, K., Yan, L., 2020. NIR-II fluorescence imaging-guided photothermal therapy with amphiphilic polypeptide nanoparticles encapsulating organic NIR-II Dye. ACS Applied Bio Materials.

Cortés, V., Blasco, J., Aleixos, N., Cubero, S., Talens, P., 2019. Monitoring strategies for quality control of agricultural products using visible and near-infrared spectroscopy. Trends in Food and Science Technology 85, 138–148.

de Aberasturi, D.J., Serrano-Montes, A.B., Liz-Marzán, L.M., 2015. Modern applications of plasmonic nanoparticles: From energy to health. Advanced Optical Materials 3, 602–617.

Delbeck, S., Vahlsing, T., Leonhardt, S., Steiner, G., Heise, H.M., 2019. Non-invasive monitoring of blood glucose using optical methods for skin spectroscopy—opportunities and recent advances. Analytical and Bioanalytical Chemistry 411, 63–77.

Demirel, G., Usta, H., Yilmaz, M., Celik, M., Alidagi, H.A., Buyukserin, F., 2012. Surface-enhanced Raman spectroscopy (SERS): An adventure from plasmonic metals to organic semiconductors as SERS platforms. Journal of Materials Chemistry 6, 5314–5335.

Different synthesis process of carbon nanomaterials for biological applications. In: Ray, S., Jana, N. (Eds.), Carbon nanomaterials for biological and medical applications. Elsevier, pp. 1–41.

Endo-Kimura, M., Kowalska, E., 2020. Plasmonic photocatalysts for microbiological applications. Catalysts 10 (8), 824.

Eustis, S., El-Sayed, M.A., 2006. Why gold nanoparticles are more precious than pretty gold: Noble metal surface plasmon resonance and its enhancement of the radiative and nonradiative properties of nanocrystals of different shapes. Chemical Society Reviews 35, 209–217.

Fan, Y., Zhang, F., 2019. A new generation of NIR-II probes: Lanthanide-based nanocrystals for bioimaging and biosensing. Advanced Optical Materials 7, 1801417.

Gastélum-Barrios, A., Soto-Zarazúa, G.M., Escamilla-García, A., Toledano-Ayala, M., Macías-Bobadilla, G., Jauregui-Vazquez, D, 2020. Optical methods based on ultraviolet, visible, and near-infrared spectra to estimate fat and protein in raw milk: A review. Sensors (Basel, Switzerland) 20 (12), 3356.

Gogoi, S., Raju Khan, R., 2018. NIR upconversion characteristics of carbon dots for selective detection of glutathione. New Journal of Chemistry 42, 6399–6407.

Grassi, S., Alamprese, C., 2018. Advances in NIR spectroscopy applied to process analytical technology in food industries. Current Opinion in Food Science 22, 17–21.

Gruber, A., Drabenstedt, A., Tietz, C., Fleury, L., Wrachtrup, J., vonBorczyskowski, C., 1997. Scanning confocal optical microscopy and magnetic resonance on single defect centers. Science 276, 2012–2014.

Guével, X.L., Hötzer, B., Jung, G., Marc Schneider, M., 2011. NIR-emitting fluorescent gold nanoclusters doped in silica nanoparticles. Journal of Materials Chemistry 21, 2974–2981.

Haedicke, K., Kozlova, D., Gräfe, S., Teichgräber, U., Epple, M., Hilger, I., 2015. Multifunctional calcium phosphate nanoparticles for combining near-infrared fluorescence imaging and photodynamic therapy. Acta Biomaterials 14, 197–207.

Harris, D.C., 2007. Quantitative chemical analysis. W.H. Freeman and Co K. Treadway (Ed.).

Harris, J.M., Martin, N.E., Modi, M., 2001. Pegylation—a novel process for modifying pharmacokinetics. Clinical Pharmacokinetics 40, 539–551.

Heller, D.A., Baik, S., Eurell, T.E., Strano, M.S., 2005. Single-walled carbon nanotube spectroscopy in live cells: Towards long-term labels and optical sensors. Advanced Materials 17 (23), 2793–2799.

Herschel, W., 1800. Experiments on the refrangibility of the invisible rays of the sun. Philosophical Transactions of the Royal Society 90, 284.

Hesse, M., Meier, H., Zeeh, B., Linden, A., Murray, M., 1997. Spectroscopic methods in organic chemistry. Universiteitsbibliotheek Gent S. Thieme (Ed.).

Hills, A.E., 2017. Spectroscopy in biotechnology research and development. In: Lindon, J.C., Tranter, G.E., Koppenaal, D.W. (Eds.), Encyclopedia of spectroscopy and spectrometry3rd ed. Academic Press, pp. 198–202.

Hollas, J.M., 2004. Modern spectroscopy. John Wiley & Sons, Ltd. E. W. Castner Jr (Ed.).

Jin, Y., Ye, F., Zeigler, M., Wu, C., Chiu, D.T., 2011. Near-infrared fluorescent dye-doped semiconducting polymer dots. ACS Nano 5 (2), 1468–1475.

Karoui, R., Downey, G., Blecker, C., 2010. Mid-infrared spectroscopy coupled with chemometrics: A tool for the analysis of intact food systems and the exploration of their molecular structure-quality relationships. Chemistry Reviews 110, 6144–6168.

Kessoku, S., Maruo, K., Okawa, S., Masamoto, K., Yamada, Y., 2011. Influence of blood glucose level on the scattering coefficient of the skin in near-infrared spectroscopy, Proceedings of AJTEC.

Kim, S., Lee, B., Lee, H., Jo, S., Kim, H., Won, Y.-Y., Lee, J., 2017. Near-infrared plasmonic assemblies of gold nanoparticles with multimodal function for targeted cancer theragnosis. Scientific Reports 7, 17327.

Kim, S., Lim, Y., Soltesz, E., De Grand, A., Lee, J., et al., 2004. Near-infrared fluorescent type II quantum dots for sentinel lymph node mapping. Nature Biotechnology 22, 93–97.

Kruss, S., Landry, M., Ende, E., Lima, B., Reuel, N., Zhang, J., Nelson, J., Mu, B., Hilmer, A., Strano, M., 2014. Neurotransmitter detection using corona phase molecular recognition on fluorescent single-walled carbon nanotube sensors. Journal of the American Chemistry Society 136 (2), 713–724.

Lalancette-Hébert, M., Moquin, A., Choi, A., Jasna Kriz, J., Maysinger, D, 2010. Lipopolysaccharide-QD micelles induce marked induction of TLR2 and lipid droplet accumulation in olfactory bulb microglia. Molecular Pharmacy 7 (4), 1183–1194.

Landau, L.D., Lifshitz, E.M., 1976. Mechanics. Pergamon Press.

Lee, C.H., Cheng, S.-H., Wang, Y.-J., Chen, Y.-C., Chen, N.-T., Souris, J., Chen, C.-T., Mou, C.-Y., Yang, C.-S., Lo, L.-W., 2009. Near-infrared mesoporous silica nanoparticles for optical imaging: Characterization and in vivo biodistribution. Advanced Functional Materials 19, 215–222.

Lee, J.C., Hwang, H.S., Lee, M.G., Kim, M., Lee, H.J., 2012. Akari near-infrared spectroscopy of luminous infrared galaxies. Astrophysics Journal 756, 1. https://doi.org/10.1088/0004-637X/756/1/95.

Lee, S., Yabuuchi, N., Gallant, B., Chen, S., Kim, B., Hammond, T., Shao-Horn, Y., 2010. High-power lithium batteries from functionalized carbon-nanotube electrodes. Nature Nanotechnology 5, 531–537.

Li, H., Shi, L., Sun, D., Li, P., Liu, Z., 2016. Fluorescence resonance energy transfer biosensor between upconverting nanoparticles and palladium nanoparticles for ultrasensitive CEA detection. Biosensors and Bioelectronics 86, 791–798.

Li, R., Tu, W., Wang, H., Dai, Z., 2018. Near-infrared light excited and localized surface plasmon resonance-enhanced photoelectrochemical biosensing platform for cell analysis. Analytical Chemistry 90 (15), 9403–9409.

Liu, T.M., Conde, J., Lipiński, T., Bednarkiewicz Huang, C.-C., 2016. Revisiting the classification of NIR-absorbing/emitting nanomaterials for in vivo bio-applications. NPG Asia Materials 8, e295.

Maas van der, J.H., 1972. Basic infrared spectroscopy. Heyden & Son Ltd.

Manley, M., 2014. Near-infrared spectroscopy and hyperspectral imaging: Non-destructive analysis of biological materials. Chemical Society Reviews 43, 8200–8214.

Martens, H., Naes, T., 1996. Multivariate calibration. John Wiley & Sons J. Sanchez (Ed.).

Martynenko, I.V., Litvin, A.P., Purcell-Milton, F., Baranov, A.V., Fedorov, A.V., Gun'ko, Y.K., 2017. Application of semiconductor quantum dots in bioimaging and biosensing. Journal of Materials Chemistry B 5, 6701–6727.

Massart, D.L., Vandeginste, B.G.M., Deming, S.M., Michotte, Y., Kaufmann, L., 2003. Chemometrics: A textbook. Elsevier.

Massey, M., Miao Wu, M., Conroy, E.M., Algar, W.R., 2015. Mind your P's and Q's: The coming of age of semiconducting polymer dots and semiconductor quantum dots in biological applications. Current Opinion in Biotechnology 34, 30–40.

Mazrad, Z., Lee, K., Chae, A., In, I., Lee, H., Park, S.Y., 2018. Progress in internal/external stimuli responsive fluorescent carbon nanoparticles for theranostic and sensing applications. Journal of Materials Chemistry B 6, 1149–1178.

Michalet, X., Pinaud, F., Lacoste, T., Dahan, M., Bruchez, M.P., Alivisatos, P., Shimon Weiss, S., 2001. Properties of fluorescent semiconductor nanocrystals and their application to biological labeling. Single Molecules 2, 261–276.

Muddana, H.S., Morgan, T.T., Adair, J.H., Butler, P.J., 2009. Photophysics of Cy3-encapsulated calcium phosphate nanoparticles. Nano Letters 9, 1559–1566.

Muhammed, M., Döblinger, M., Rodríguez-Fernández, J., 2015. Switching plasmons: Gold nanorod–copper chalcogenide core–shell nanoparticle clusters with selectable metal/

semiconductor NIR plasmon resonances. Journal of the American Chemistry Society 137 (36), 11666–11677.

Ning, Y., Zhu, M., Zhang, J.-L., 2019. Near-infrared (NIR) lanthanide molecular probes for bioimaging and biosensing. Coordination Chemistry Reviews 399, 213028.

Olesberg, J.T., Liu, L., Van Zee, V., Arnold, M.A., 2006. In vivo near-infrared spectroscopy of rat skin tissue with varying blood glucose levels. Analytical Chemistry 78, 215–223.

Ozaki, Y., 2012. Near-infrared spectroscopy—its versatility in analytical chemistry. Analytical Sciences 28 (6), 545–563.

Pal, K., Sharma, V., Sahoo, D., Kapuria, N., Koner, A.L., 2018. Large Stokes-shifted NIR-emission from nanospace-induced aggregation of perylenemonoimide-doped polymer nanoparticles: Imaging of folate receptor expression. Chemical Communications 54, 523.

Pan, J., Li, F., Choi, J.H., 2017. Single-walled carbon nanotubes as optical probes for bio-sensing and imaging. Journal of Materials Chemistry B 5, 6511–6522.

Park, J., Dvoracek, C., Lee, K.H., Galloway, J., Bhang, H., Pomper, M.G., Searson, P., 2011. CuInSe/ZnS core/shell NIR quantum dots for biomedical imaging. Small 7 (22), 3148–3152.

Pasquini, C., 2003. Near infrared spectroscopy: fundamentals, practical aspects and analytical applications. Journal of Brazilian Chemical Society 14 (2), 198–219.

Pasquini, C., 2018. Near infrared spectroscopy: A mature analytical technique with new perspectives—a review. Analytica Chimica Acta 1026, 8–36.

Pavia, D., Lampman, G.G.K., 2001. Introduction to spectroscopy: A guide for students of organic chemistry. Harcourt College Publishers.

Perkin Elmer Life and Analytical Sciences. (2005). FT-IR spectroscopy-attenuated total reflectance (ATR). https://cmdis.rpi.edu/sites/default/files/ATR_FTIR.pdf.

Perkins, W.D., 1986. Fourier transform-infrared spectroscopy: Part l. Instrumentation. Journal of Chemical Education 63 (1), A5.

Petryayeva, E., Algar, W., 2013. Proteolytic assays on quantum-dot-modified paper substrates using simple optical readout platforms. Analytical Chemistry 85 (18), 8817–8825.

Petryayeva, E., Algar, W., 2014. Multiplexed homogeneous assays of proteolytic activity using a smartphone and quantum dots. Analytical Chemistry 86 (6), 3195–3202.

Prati, S., Sciutto, G., Bonacini, I., Mazzeo, R., 2016. New frontiers in application of FTIR microscopy for characterization of cultural heritage materials. Topics in Current Chemistry 374 (3), 26.

Rahman, M., Mieno, T., 2015. Conductive cotton textile from safely functionalized carbon nanotubes. Journal of Nanomaterials, 978484.

Rijeka, C., 2012. Introduction to infrared spectroscopy. Infrared Spectroscopy - Materials Science, Engineering and Technology. InTech T. Theophanides (Ed.).

Roggo, Y., Chalus, P., Maurer, L., Lema-Martinez, C., Edmond, A., Jent, N., 2007. A review of near infrared spectroscopy and chemometrics in pharmaceutical technologies. Journal of Pharmaceutical Biomedical Analytics 44, 683–700.

Sakudo, A., 2016. Near-infrared spectroscopy for medical applications: Current status and future perspectives. Analytica Chimica Acta 455, 181–188.

Santra, S., Dutta, D., 2007. Nanoparticles for optical imaging of cancer. In: Kumar, C.S. (Ed.), Nanomaterials for cancer diagnosis. Wiley-VCH Verlag GmbH, pp. 44–85.

Sargent, E.H., 2005. Infrared quantum dots. Advanced Materials 17 (5), 515–522.

Schrader, B., 1995. Infrared and Raman spectroscopy: Methods and applications. VCH Verlag.

Singh, A., Mukherjee, M., 2005. Effect of polymer-particle interaction in swelling dynamics of ultrathink nanocomposite films. Macromolecules 38, 8795–8802.

Skoog, D.A., Holler, F.J., Crouch, S.R., 2007Principles of instrumental analysis6. Brooks/Cole, Cengage Learning Edited by M. A. Belmont.

Soppimath, K.S., Aminabhavi, T.M., Kulkarni, A.R., Rudzinski, W.E., 2001. Biodegradable polymeric nanoparticles as drug delivery devices. Journal of Controlled Release 70, 1–20.

Sotiriou, G.A., 2013. Biomedical applications of multifunctional plasmonic nanoparticles. Wiley Interdisciplinary Review of Nanomedicine and Nanobiotechnology 5, 19–30.

Tan, L., Wan, J., Guo, W., Ou, C., Liu, T., Fu, C., Zhang, Q., Ren, X., Liang, X.-J, Ren, J., Li, L., Meng, X., 2018. Renal-clearable quaternary chalcogenide nanocrystal for photoacoustic/magnetic resonance imaging guided tumor photothermal therapy. Biomaterials 159, 108–118.

Tran, C.D., 2005. Principles, instrumentation, and applications of infrared multispectral imaging, an overview. Analytical Chemistry Letters 38 (5), 735–752.

Tycko, B., Maxfield, F.R., 1982. Rapid acidification of endocytic vesicles containing α2-macroglobulin. Cell 28, 643–651.

Urries, I., Muñoz, C., Gomez, L., Marquina, C., Sebastian, V., Arruebo, M., Santamaria, J., 2014. Magneto-plasmonic nanoparticles as theranostic platforms for magnetic resonance imaging, drug delivery and NIR hyperthermia applications. Nanoscale 6, 9230–9240.

Valizadeh, A., Mikaeili, H., Samiei, M., Farkhani, S., Zarghami, N., Kouhi, M., Akbarzadeh, A., Davaran, S., 2012. Quantum dots: Synthesis, bioapplications, and toxicity. Nanoscale Research Letters 7, 480.

Wang, H., Mukherjee, S., Yi, J., Banerjee, P., Chen, Q., Zhou, S., 2017. Biocompatible chitosan–carbon dot hybrid nanogels for NIR-imaging-guided synergistic photothermal–chemotherapy. ACS Applied Materials Interfaces 9 (22), 18639–18649.

Wang, Y., Xia, X., Wang, X., Ku, G., Gill, K.L., O'Neal, P., Stoica, G., Wang, L.V., 2004. Photoacoustic tomography of a nanoshell contrast agent in the in vivo rat brain. Nano Letters 4, 1689–1692.

Wold, J., Johansen, I.-R., Haugholt, K., Tschudi, J., Thielemann, J., Segtnan, V., … Wold, E., 2006. Non-contact transflectance near infrared imaging for representative on-line sampling of dried salted coalfish (bacalao). Journal of Near Infrared Spectroscopy 14.

Yang, W.L., Nair, D.G., Makizumi, R., Gallos, G., Ye, X., Sharma, R.R., Ravikumar, T.S., 2004. Heat shock protein 70 is induced in mouse human colon xenografts after sublethal radiofrequency ablation. Annals of Surgical Oncology 11, 399–406.

Yang, Y., Nie, C., Deng, Y., Cheng, C., He, C., Ma., L., Zhao, C, 2016. Improved antifouling and antimicrobial efficiency of ultrafiltration membranes with functional carbon nanotubes. RSC Advances 6, 88265–88276.

Zaytseva, O., Neumann, G., 2016. Carbon nanomaterials: Production, impact on plant development, agricultural and environmental applications. Chemical and Biological Technology in Agriculture 3, 17.

Zhang, J., Shikha, S., Mei, Q., Liu, J., Zhang, Y., 2019. Fluorescent microbeads for point-of-care testing: A review. Microchimica Acta 186, 361.

Zhang, W., Patel, K., Schexnider, A., Banu, S., Radadia, A., 2014. Nanostructuring of biosensing electrodes with nanodiamonds for antibody immobilization. ACS Nano 8 (2), 1419–1428.

Zhao, J., Zhong, D., Zhou, S., 2018. NIR-I-to-NIR-II fluorescent nanomaterials for biomedical imaging and cancer therapy. Journal of Materials Chemistry B 6, 349–365.

Zhao, P., Xu, Q., Tao, J., Jin, Z., Pan, Y., Yu, C., Yu, Z., 2018. Near infrared quantum dots in biomedical applications: Current status and future perspective. Wiley Interdisciplinary Reviews in Nanomedicine and Nanobiotechnology 10, e1483.

Advances in microscopy and their applications in biomedical research

11

Sonali Karhana[a], Madhusudan Bhat[a], Anupama Ninawe[b], Amit Kumar Dinda[a]
[a]*Department of Pathology, All India Institute of Medical Sciences, New Delhi, India*
[b]*Department of Biochemistry, All India Institute of Medical Sciences, New Delhi, India*

11.1 Microscopy as an integral part of biomedical research

The ability to view and analyze specimens under high resolution has always been crucial for biomedical sciences. Various imaging modalities are used for viewing and analysis of biologic specimen, which includes imaging in both clinical and research settings. Clinical modalities consist of ultrasound, X-ray, computed tomography (CT), magnetic resonance imaging (MRI), endoscopy, and optical coherence tomography (OCT), among others, while electron microscopy, mass spectrometry imaging, and various optical microscopy techniques (e.g., confocal, total internal reflection, multiphoton, superresolution fluorescence microscopy) are included in research modalities. A combination of both clinical imaging and research microscopy is required for interpretation of clinical datasets. Imagining a world without imaging is now impossible with the developments in translational studies and medicine with the advent of better microscopes. As new technologies are gaining weight with time, researchers and medical practitioners are now being able to gain more complimentary information, thereby improving the prognosis, follow-up, and treatment of various diseases.

Combination of engineering sciences, chemical engineering, and computational studies together continue to contribute toward the advancement of microscopy in biomedical research. Imaging in research has gained tremendously with the development of engineering sciences that have provided better detectors, high sensitivity, increased speed, better system designs, and enhanced resolution. Its contribution to biomedical sciences and translational studies includes single cell imaging, biomaker detection, imaging therapeutics, among others. Apart from advancements in detector and optics technologies that are contributed by engineering sciences, chemical engineering has provided fluorescent probes for imaging biologic specimens (Livet et al., 2007). These chemical probes offer cellular and molecular specificity that has allowed scientists to visualize specific biomolecules and cells. Reporter genes have also been developed for MRI (Weissleder et al., 2000), ultrasound (Shapiro et al., 2014), and nuclear imaging (Gambhir et al., 2000). Additionally, computational advancement has provided automated image analysis, integration of complex datasets to multiscale models, and data mining. Since the invention of the first microscope in the 16th century, various types of microscopes have been developed, which can be broadly categorized according to the illumination source used for imaging, namely electrons and photons. The microscopy techniques that employ photons as their illumination source are grouped under the category of photon-based imaging, while the microscopy techniques that use electrons for illuminating the specimens are grouped under electron-based microscopy.

11.2 Photon-based imaging

Photon-based microscopy includes techniques that involve photons as their illumination source, and these microscopes are known as optical microscopes or light microscopes. Photon-based microscopic imaging modalities used in biomedical research mainly include bright field microscope, dark field microscope, phase contrast microscope, differential interference contrast microscope, confocal microscope, fluorescence microscope, near-field scanning optical microscopy (NSOM), multiphoton microscopy, among others. These modalities have the ability to provide an image of a given sample of tissue at cellular level. They also provide adjustable magnification and depth of field that ensures a well-focused image.

In **bright field microscopy**, the source of light is situated at the bottom, which illuminates the specimen placed above, so the image depends upon the light being transmitted by the specimen. The specimen to be observed absorbs part of the light and transmits the remaining through the ocular lens as a result of which the specimen appears darker than the surrounding field. The area around the specimen is brightly illuminated, thus this technique's name. This type of microscopy is generally used to study the morphology of cells, but it requires staining to increase the contrast of the image. Stains can be generally divided into two types: basic dyes having positive charge and acidic dyes that are negatively charged. Cell components exhibiting negative charge, like nucleic acids bind to basic dyes. Positively charged components, in turn, are attracted to acidic dyes. The positively charged stains (i.e., basic dye) attach promptly to the specimen and are used more often than the acidic dyes as negative stains are readily repelled by the negative charge on the surface of the cells. Acidic dyes are often employed to stain surrounding field to increase contrast. Some of the widely used positive stains are methylene blue and Rose bengal. These stain the whole cell, which can be visualized against a bright unstained background. Differential stains involve two stains, namely primary stain and counter stain. The most widely used differential staining was developed by Hans Christian Gram, known as Gram staining. It utilizes two stains, crystal violet and safranin. The difference in staining occurs due to the difference in the components of the cell walls of gram-positive and gram-negative bacteria. For staining bacterial spores, capsules, or species like *Mycobacterium* sp., Gram staining cannot be used, so acid-fast staining is employed. These bacteria cannot be destained when treated with carbolfuchsin and are also known as acid-fast bacteria.

In an attempt to increase the contrast and resolution of the standard bright field microscope, several techniques have been developed with slight differences (e.g., dark field microscopy, phase contrast microscopy, and differential interference contrast [DIC]).

Dark field microscopy has only slight changes to bright field microscopy. In this type of microscope, a central stop is attached just above the condenser lens, which allows only a limited amount of light to reach the objective lens. Therefore only the light scattered through the edges of the specimen is used to view the image. In this way the surrounding field is obtained dark and the specimen appears bright. This technique ensures better contrast than the standard bright field microscopy. Dark field microscopes are generally used to study morphology and motility of cells.

Phase contrast microscopy is another attempt to increase the contrast of the transparent samples. This technique is based on the concept that all the transparent cellular components have different densities. Since light interacts differently with varying amounts of densities, a contrast is observed between the cellular components and the outside surrounding area. To focus the light wave on single focal plane, Kohler illumination is put into use. The light is focused onto the specimen as hollow cone. A diffraction plate is situated at the back focal point of the objective lens. The diffraction plate alters the phase of the ray of light that comes through the specimen. This results in varying degrees of retardation that causes either darkening or lightening of the sample to be viewed.

Differential interference contrast microscopy gives three-dimensional (3D), colored, and highly contrasting images. In this technique the light ray that will illuminate the specimen is divided into two different beams of light. The first beam reaches the specimen and passes through it, while the other beam does not. This results in a phase difference between the two beams of light. Both the beams of light are then combined to cause interference. With such images, we can detect minor elevations or variations in depth in the specimen that gives the perception of 3D image. The abovementioned types of photon-based microscopes have their fair share of contribution majorly involving analysis of morphology of cells. Apart from these, two optical microscopes have stood out and made exceptional contributions to the field of biomedical sciences: fluorescence microscope and confocal microscope.

11.2.1 Fluorescence microscopy: Working, advancements, and biomedical applications

Fluorescence microscopy has become a major tool to analyze cell physiology. The phenomenon of absorption and subsequent reemission of photons, formally called fluorescence, has had a major impact on biomedical research as it not only enables visualization with good spatial resolution but also allows researchers to track single molecules in vivo and replace microscopy to age-old radioactive assays. These microscopes offer better contrast, resolution, and variety in design when compared with the standard light microscopes.

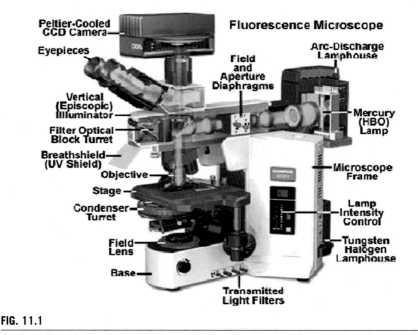

FIG. 11.1

Configuration of a fluorescence microscope.

The excitation light from a suitable light source is made to focus on the specimen via the objective lens. The emitted fluorescence coming from the specimen is directed toward the detector. Fig. 11.1 explains the trajectory of light rays in a typical fluorescence microscope. Fluorescence microscopy involves the use of various fluorescent dyes such as fluorescein or acridine orange. These dyes are used to stain the specimen, which has to be observed under fluorescence microscope. The underlying principle involved in using such dyes is Stokes shift. When photon of a specific wavelength hits the fluorescent molecule, an electron is transferred from its ground state to excited state. Electron cannot stay in the excitation band for long as it is in a very unstable state, so it travels back to its ground state in just a matter of some nanoseconds. While traveling back, a photon with lesser energy than the incident photon is released, which is detected as fluorescence. Therefore the wavelength of light emitted from the fluorescent dye is always higher than the wavelength of light incident on the stain. Since a major part of the excitation light is transmitted through the specimen, the main aim of fluorescence microscopy is to separate the excitation light from the emitted light. Therefore indicators or fluorescence dyes that exhibit large Stokes shifts are preferred. This separation of the bright excitation light from the comparatively dimmer emitted light is done by using optical filters. The fundamental key to obtaining a good image lies with choosing the right filter. Therefore the choice of filter is a compromise between emitted light (passing through) and excitation light (blocking rays).

11.2.1.1 *Advancements and biomedical applications of fluorescence microscopy*

New advancements have been introduced to fluorescence microscopy, which include employing photobleaching-based techniques like fluorescence recovery after photobleaching (FRAP), fluorescence localization after photobleaching (FLAP), Förster or fluorescence resonance energy transfer (FRET), and the related fluorescence loss in photobleaching (FLIP).

FRAP was first introduced while studying the rate of fluorescence recovery at an already bleached site (Axelrod et al., 1976; Koppel et al., 1976). A typical FRAP experiment involves irreversible photobleaching of fluorescent molecules in a particular area of the cell by using a focused laser beam at a high intensity. This results in diffusion of the nonphotobleached fluorescent molecules to the site where laser was applied, which in turn facilitates recovery of fluorescence at a particular rate. The rate of this fluorescence recovery is then recorded. Initially FRAP was applied to analyze the rate of diffusion through cell membranes by using fluorescein or any other organic dyes (Liebman & Entine, 1974). With time, FRAP started to be used for studying protein mobility in the interior of the cell. Now, with the advent of new genetic engineering techniques allowing researchers to make recombinant proteins, it is possible to perform imaging with multiple tags. Therefore FRAP has evolved from the study of diffusion of plasma membrane phospholipids and proteins (Jacobson et al., 1987; Peters & Beck, 1983; Tocanne et al., 1989) to analysis of protein dynamics within and among cells (Kimura et al., 2003; Reits & Neefjes, 2001; White & Stelzer, 1999). Moreover, FRAP has also been shown to be a technology of choice for the analysis of nuclear protein dynamics in live cells (Monetta et al., 2007).

Further, a new modification to FRAP has been introduced, which is known as inverse FRAP (iFRAP). In this method, all the constituent fluorochromes of the cell are bleached except the fluorochromes of a selected part of an organelle. The rate of loss of fluorescence of the selected area is then recorded. This gives the rate of exchange of molecules from the surroundings to the organelle. This methodology has been used to analyze the mobility of molecules in the nucleus and their exchange with the surrounding cytoplasm (Dundr et al., 2002).

FLIP is an alternative technique to FRAP, wherein, unlike FRAP and iFRAP experiments, bleaching is done repeatedly at the same region of the specimen, thereby not allowing fluorescence recovery. This repeated bleaching is done around the unbleached region of interest. FLIP experiments have known to be useful for the demonstration of flux and connectivity among various regions of cell. Thus it is an ideal method to study and analyze exchange of molecules between compartments of the cell. FLIP has been used to observe the protein traffic between the nucleus and the cytoplasm and in endoplasmic reticulum (Nehls et al., 2000; Phair & Misteli, 2000).

FLAP is a technique devised to overcome some of the limitations of FRAP and FLIP. The tracking of unlabeled molecules is not possible using FRAP and FLIP. To overcome this limitation, FLAP involves fluorescent labeling of the selected biomolecules twice. First fluorescent label is bleached, and the second remains intact. The second label is used as a reference label. The absolute FLAP signal is acquired by subtracting the bleached signal from the unbleached signal, which facilitates the tracking of the biomolecules. This technique can distinguish biomolecules that move at different speeds. Using FLAP, a team was able to demonstrate that monomeric G-actin protein relocates faster than filamentous F-actin (Dunn et al., 2002).

FRET is based on energy transfer (nonradiatively) from an excited donor to another biomolecule or acceptor. The acceptor molecule does not need to be fluorescent. FRET results rely on the closeness or proximity of the donor and acceptor biomolecules and cannot function if their proximity is more than 10 nm. Therefore FRET can be used to determine the molecular proximity and interactions between adjacent molecules, which lie below the resolution range of a standard optical microscope. Organic fluorescent dyes and fluorescent proteins have proved to be successful in setting up FRET experiments. Green fluorescent protein (GFP)–based FRET imaging has been used to determine the functional organization and compartmentalization of living cells as well as tracing the mobility of proteins within and among cells (Jares-Erijman & Jovin, 2003). FRET-based biosensors have also been devised, of which the most popular one is Cameleon, a family of Ca^{2+} sensors based on use of calmodulin for sensing (Miyawaki et al., 1997).

In addition to introduction of new techniques, advancements in various tools related to fluorescence microscopy have also resulted in obtaining high-quality images such as decrease in signal-to-noise ratio and attaching ICCD and EMCCD cameras (Petty, 2007). Many such modifications to advanced fluorescence microscopy techniques are still underway, and surely the future is bright. Microbiologists use fluorescence microscopy for performing direct counts of the microorganisms, the most widely used stain being acridine orange direct counts (AODC). One of the most noteworthy use of fluorescence microscopy is the detection of certain protein molecules (antibodies) in the case of immunolabeling or specific nucleic acid sequences, called in situ hybridization, by employing probes attached with fluorescent molecules. Fluorescence microscopy has been applied to study the dental hard tissue as well. The tooth crown of humans, dentine is layered by enamel and the tooth root is layered by cementum. Cementum is more porous and softer than the enamel. On exposure of the tooth to the oral environment, caries can occur readily. These changes in dentine due to caries affect the optical properties of dentine. Thus fluorescence measurements can be used for examining carious portions (Kvaal & Solheim, 1989; Sundström et al., 1985).

Although most of the techniques mentioned are technically demanding, they are gaining acceptance in the biomedical community at a good pace and soon can be a possible mandatory item in a researcher's tookit.

11.2.2 Confocal microscopy: Working, advancements, and biomedical applications

Confocal microscopy provides high-resolution images of thin sections of the specimen both ex vivo and in vivo. This technique allows the user to obtain sectioned images of normal tissue and lesions for their structural and morphologic analysis.

These structural features are then used to differentiate between normal and abnormal cells. Conventional microscopes work on the principle of Kohler illumination, while the inventor of confocal microscope, Marvin Minsky, introduced the concept of point illumination. Kohler illumination offers a wide field, which allows the out-of-focus light to hamper with the imaging. In the case of point illumination, a single point on the focal plane is illuminated. This is achieved by keeping the focal point of both the objective and condenser lenses the same, hence the name *confocal*. This arrangement of lenses ensures a single point focus and imaging (Watson, 1997). A pinhole is located at the focus point right above the detector, which allows only the in-focus rays to reach the detector. The light rays are focused on a small area of the specimen, and the condenser lens (having the same focal length as the objective lens) is focused on the same area. Thus the imaging is done only through the light reflected by the focal plane as adjusted by the user. Such arrangement of the lenses and subsequent focusing offers better resolution than the conventional light microscope (Fig. 11.2). This kind of optical sectioning facilitates imaging the interiors of both the translucent and opaque biologic specimens. A laser is employed to scan the specimen one slice and one pixel at a time in confocal laser scanning microscopy (CLSM), which is used to obtain a digital image.

The whole image construction is done by scanning the specimen. Three types of scanning designs have been set up to date, namely stage scanning, beam scanning, and tandem scanning (Wright & Schatten, 1991). Overview of the three scanning designs is summarized in Table 11.1.

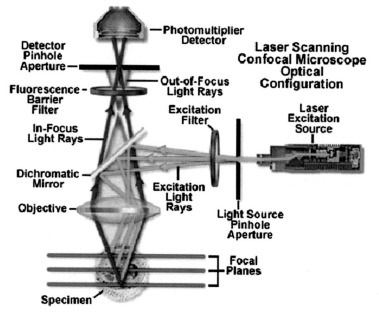

FIG. 11.2

Working of a confocal microscope.

Table 11.1 Difference between the three scanning designs offered in confocal microscopes.

Scanning design	Specifications	Advantages	Limitations
Stage Scanning	The stage moves under a stationary light beam.	Larger field of view Less optical aberrations from condenser and objective	Low scanning speed The specimen has to be fixed firmly to the stage to prevent blurred images Cannot perform in vivo observations
Beam Scanning	The light beam moves over a stationary stage.	Scanning speed allows for video-rate clicks	The use of galvanometer causes mechanical limitations
Tandem Scanning	Both the stage and light beam are allowed to move. This design employs thousands of apertures drilled into a glass-metal or a ceramic-copper disk instead of a single pinhole.	Provides real-time imaging	Requires skilled professionals and is expernsive.

Light sources employed for confocal microscopy can be both monochromatic and polychromatic. Tungsten, xenon, mercury, carbon, quartz-halogen, and zirconium lamps that emit white light, which includes light rays of many wavelengths, are known as polychromatic light sources (Chen et al., 1990). These light sources are easily available, cheap, and offer wavelength flexibility when used with various wavelength filters. These are widely used for tandem scanning confocal microscopy. Monochromatic light sources include lasers, which were first used for confocal microscopy in 1969 (Davidovits & Egger, 1969). Lasers exhibit some unique properties, which makes them a better alternative over polychromatic light sources, such as high coherence, and low degree of divergence which allows the user to focus the beam at a point, high brightness allows for better and clearer image, and high degree of monochromaticity enables the work to be done at a specific wavelength (Gratton & VandeVan, 1989).

11.2.2.1 *Advancements and biomedical applications of confocal microscopy*

Advancement in the field of computer sciences has contributed to the development of digital image processing, and this has been majorly resourceful for confocal microscopy to develop over the years (Petroll et al., 1992). Confocal microscopes majorly employ software for image acquisition and further image processing.

Images can be processed both during image acquisition and post image acquisition (Wright et al., 1993). To date, several image processing methodologies are available, all of which generally include three procedures: image manipulation, display, and analysis (Chen et al., 1990). Image manipulations involve contrast enhancement, smoothing, image alignment, etc. But the most widely used image manipulation is the 3D reconstruction. It stacks the optical sections acquired by focusing on the specimen at different levels (Wilson, 1989). Image analysis can be carried out using numerous operations like calculation of distances, angles, volumes, and surfaces. The collected information can then in turn be projected in several ways: rotation of object, codification and coloring of different light intensities, choice of background color, projection on a specific surface, dye labeling, among others (Wright & Schatten, 1991). Broadly, there are three major ways to acquire an image using confocal microscopy. First is the 3D reconstruction that has proved to be immensely helpful in obtaining images of thick specimens. The second method involves representing the various optical sections as z-series montage at regular intervals. The third methodology involves providing two images as viewed from two different angles of the specimen, which results in an illusion of a 3D image (Boyde, 1985; Wright et al., 1993).

Various fields of life sciences have benefited from the quality of images provided by confocal microscopy. It is mostly considered as a better alternative to biopsy and subsequent conventional microscopy analysis. For instance, detection of melanoma involves histologic examination of lesions, but the whole procedure requires biopsies and can cause unnecessary sampling errors or distortion of tissue. The procedure is time consuming, and dynamic observation with time cannot be performed. Confocal microscopy, on the other hand, offers noninvasive slice-by-slice sectioning of tissue and does not require slide preparation as in the case with normal bright field microscopic observations.

Confocal microscopy has been widely applied for investigative analysis of skin as the depth range of the instrument is around 200 to 300 μm, which allows for viewing the different layers of the skin. Along with the different layers of skin, the user is able to obtain images of sebaceous glands, sweat ducts, and hair follicles, which easily correlate with histology. Use of confocal microscopy ensures that the method remains painless for the patients, and it is feasible to go for multiple rounds of imaging. Morphometric variations between the sun-exposed and sun-protected regions of the skin have been observed using confocal microscopy (Huzaira et al., 2001). It has also been used to visualize keratinocytes, spinous cells, basal keratinocytes, circulating blood cells, extracellular matrix fiber, and corneocytes (Selkin et al., 2001). Different constituents of the skin have different refractive indices, such as melanin, and cytoplasm reflects different intensity light rays to the detector. Therefore homogeneous layers like stratum corneum exhibit low contrast and thus look darker, meanwhile heterogeneous layers like viable epidermis exhibit higher contrast and look brighter. This causes a clearly visible well-defined junction between the stratum corneum and viable epidermis (Caspers et al., 2003). Apart from the wide usability shown in dermatology examinations, confocal microscopy has proved to be useful in the study of fluorescent transport in brain capillaries (Miller et al., 2000) and for the analysis of pancreatic islet composition and architecture (Brissova et al., 2005). An accuracy of 89.7% of confocal microscopy has been reported for detecting cancerous growth in colon, esophagus, and stomach (Inoue et al., 2000).

While confocal microscopy has found its usage majorly for immediate biopsy, the real advantage that it holds is its property to noninvasively image a specimen tissue, especially that of gastrointestinal tract, eye, and the skin. This proves to be gainful in procedures like guided surgeries. It gives immediate observations of probable pathologies as well. For instance, in a case where premalignant and benign lesions in the oral cavity could not be clearly distinguished, a laser confocal endomicroscopy was used in vivo, which revealed the difference. The diagnosis confirmed that it can be used to point out structural differences between lesion and normal tissue of the oral cavity (Thong et al., 2007). Confocal microscopy has also proved to be useful in detecting differences in morphologies of proliferative skin lesions, malignant pigmented skin lesions, and nonmelanoma skin cancer (Selkin et al., 2001).

Since the advent of use of lasers for diagnostics, a variety of types of lasers have been developed and studied. Different types of lasers were administered to rhesus monkeys to develop laser-eye level thresholds (Sliney & Wolbarsht, 1980; Winburn, 1985). Consequently, lasers were divided into four groups according to their level of harmfulness. Several factors control the effect of laser on the human body like the power of laser, type, and exposure time. Thus, with these three parameters controlled, confocal microscopy can be used in human clinical studies while keeping in mind all the safety precautions.

Confocal microscopy has also been useful for studying drug delivery by combining it with fluorescence techniques (White & Errington, 2005). For instance, a pH-sensitive fluorescence dye was used to visualize and measure the surrounding pH during the time of degradation of poly(lactic-co-glycolic acid) (PLGA) microspheres, which is primarily a controlled drug release system (Pygall et al. (2007)] Similarly, CLSM has been used to analyze retention of a polymer on mucosal membrane [Batchelor et al. (2004)]. Characterization of pharmaceutical formulations like microspheres, pellets, colloidal systems, hydrophillic matrices and film coatings have been possible with the use of Confocal Microscopy (Pygall et al., 2007).

11.3 Automated imaging workstations for optical microscopy

The methodologies and subsequent advancements of optical microscopy are maturing at a rapid rate. The next step to add on to this growing field is to improve the image quality, so as to perform quantitative analysis. This problem has led to developments in image processing known as automated image processing systems

(AIPS). It consists of both hardware and software that together are used for recording, display processing, and storage of images (Altun & Taghiyev, 2017). Combining optical microscopy techniques with automated image processing and analysis tools enables the researchers to move a step beyond the conventional 2D qualitative analysis. Standard light microscopy deals with the limitations of contrast, resolution, image-to-noise ratio, and sampling statistics in 3D. Confocal optics and image deconvolution help optimize all these parameters and thus provide better images and increase the range of specimens that can be observed. These tools can especially be used for specimens that are thicker than the depth of field of the microscope. For acquiring 3D images, large data sets have to be generally processed. These data sets are difficult to analyze directly. This problem is solved by automated montaging (i.e., the computational designing of large images from a number of overlapping partial views). The resulting image is called a mosaic or a montage. A lot of work is happening in the development of computer algorithms to get as resolved images as possible. These methods include handling biologic variability and noise, since the data obtained by biologic imaging show a high degree of variability, and coping with it is most essential especially when image analysis is being performed for unsupervised operation for a large data set (Roysam et al., 1992). Some classical problems in both medicine and biology are the determination of the effect of a drug, an electrochemical stimuli, a toxin, or a state of a disease on a particular tissue. The impact can be measured in the form of structural changes, deviations in cell count, ploidy distribution, volume, or tissue architecture. Detection of such subtle changes requires comparisons at greater details. Current methods in conjugation with automated 3D imaging and subsequent image analysis offer multiple advantages. These methods allow for rapid sampling since only smaller thick samples are required as specimens. It also provides much spatial information that is generally lost when physical sectioning is performed for current techniques. Automated image analysis also provides sampling of high number of specimens and is cheaper than the current methods. The most sought-after advantage of computer automated imaging might be its propensity to not deliver human-generated errors (Turner et al., 1997). This prevents inconsistency of the generated data, which surely is highly desirable for research studying small molecules.

Recently, a research group was able to successfully segment the brain region automatically to obtain micro-optical images with the help of the convolutional neural network (CNN). This promises to be a valuable tool for neuroscientists worldwide (Tan et al., 2020). New studies have shown that artificial intelligence (AI) has the ability to change one kind of contrast to another. This property was used by a group of researchers who applied it to fluorescence microscopy as an imaging tool for cell biology. They presented phase imaging with computational specificity (PICS), which facilitated imaging of unlabeled cells, too (Kandel et al., 2020). Automated imaging workstations have revolutionized the way we garner imaging in biosciences. With the introduction of AI and CNN, imaging systems are sure to progress to greater heights in the near future.

11.4 Near-field scanning optical microscopy (NSOM): Principle and biomedical applications

Traditionally, optical microscopy has been most widely employed for imaging of biomedical samples pertaining to their noninvasiveness, ease of use, and specificity. But the attainable spatial resolution is limited to almost half of the wavelength of light, which makes the resolution limit of 200 to 300 nm for visible radiation. This limitation drove the development of electron microscopic techniques (scanning electron microscope, transmission electron microscope) and the recently introduced scanning probe techniques (atomic force microscopy, scanning tunneling microscopy). But these higher resolution techniques demand high labor for sample preparation. Atomic force microscopy reveals topographic information but is not suitable for analysis of chemical information and interpretation of specimens with complicated topography. Therefore a researcher would want to combine high resolution, specificity, sensitivity, and flexibility. These goals have inspired the development of NSOM, which can be used to image specimens with resolution beyond the diffraction limit. The diffraction limit was described in detail in 1873 by Ernst Abbe. The maximum resolution comes out to be around 250 to 300 nm. This limitation led many researchers to explore alternative optical measurements to obtain higher resolution. In the early 20th century, Synge (1928) proposed a new kind of optical instrument designed to circumvent the disadvantages posed by the problem of diffraction limit. NSOM is based entirely on the theories proposed by Synge. He proposed devising a microscopic aperture in an opaque screen, the dimensions of which are smaller than the optical wavelength. Using a high-intensity light source and illuminating the opaque screen with it, the light passing through the hole remains confined as per the dimensions of aperture. If this setup is placed in close proximity to the specimen, then the light rays emerging from the hole can be employed to image the specimen (Fig. 11.3). Subsequently, the experimental feasibility of this technique was shown in 1972 by using microwave radiation (Ash & Nicholls, 1972). Until the mid-1980s IBM Zurich and a team from Cornell University worked on

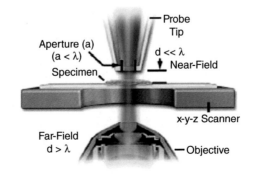

FIG. 11.3

Description of the near-field region when light is illuminated through an aperture.

overcoming the technical limitations attached with using subwavelength apertures, which finally paved the way to development of modern-day NSOM. Though still not used completely in routine, high-resolution NSOM offering precise optical measurements has begun to be used by biomedical researchers.

The potential of applications using NSOM to garner insights at a submicrometer level is expanding. NSOM has been used to probe genetic material (Garcia et al., 1998; Moers et al., 1996; Subramaniam et al., 1997), green fluorescent protein (Muramatsu at el., 1996; Tamiya et al., 1997), membrane organization in cells (de Paula et al., 2001; Talley et al., 1996), and several other biologic systems (Keller et al., 1998; Smith et al., 1993). Blood cells of humans infected with the malarial parasite *(Plasmodium falciparum)* were studied under NSOM so as to study the interaction between host proteins and parasitic proteins. Selective labeling of the two proteins (i.e., parasite and host with fluorescent dyes) was followed by imaging using NSOM (Enderle et al., 1998). As NSOM offers high resolution, it is also being applied in areas of genome mapping (Moers et al., 1996; Muramatsu et al., 1996; Subramaniam et al., 1997). NSOM has also contributed to the study of the organization of molecules within the cellular membrane. Various works have claimed the presence and functional importance of lateral heterogeneities in cell membranes (Ha et al., 1996; Tamm et al., 1996). This technology is now also being developed for imaging live cells, and work is still in the preliminary stage. Recently, NSOM was combined with atomic force microscopy to visualize protein molecules for their real-time functional activity analysis (Umakoshi et al., 2020). The new age photoconductive antenna microprobe (PCAM)–based terahertz NSOM has been applied for imaging tissue of mouse brain. It could unambiguously distinguish the cerebrum with the corpus callosum region (Geng et al., 2019). Subsequently, near-field terahertz sensing has been used to visualize HeLa cells, *Pseudomonas* sp. among other biologic samples (Bai et al., 2020). Undoubtedly this technique is still in its early stage, but equally true is the fact that it has already been developed to a point where useful and unique measurements of a wide variety of biologic samples have contributed to biomedical research.

11.5 **Electron-based imaging: Electron microscopy (EM)**

Electron microscopy delivers detailed high-resolution images of tissues, cells, cell components, and even macromolecules. The source of illuminating radiation in this type of microscopy is a beam of electrons, which exhibit a very short wavelength. The high resolving power of the microscope is because of the shorter wavelength and focusability of electron beam. EM offers a magnification up to 1,000,000X, which allows for detailed observation of very fine structures of cellular components. EMs are conceptually similar to optical microscopes with some fundamental variations listed in Table 11.2.

In an electron microscope, an electron beam is applied onto the specimen placed in a vacuum chamber. This electron beam is focused by using a series of

Table 11.2 Comparison between optical microscope and electron microscope.

Characteristic	Optical Microscope	Electron Microscope
Illuminating beam	Light beam	Electron beam
Type of specimen that can be seen	Live or dead	Dead or dried
Sample preparation	Specimen is stained with dye	Specimen is coated with heavy metals to reflect electrons
Sample preparation time	Less	More
Lens	Glass lens	Electromagnetic lens
Magnification	Up to 2000X	Up to 1,000,000X
Resolving power	Low (2000 Å)	High (3 Å)
Image	Colored	Black and white
Vacuum requirement	No	Yes
Focusing	Mechanical	Electrical
High-voltage electricity	Not required	Voltage of ≥50,000 required
Cooling system	Not required	Required, to take out heat generated by high electric current
Risk	No radiation risk	Risk of radiation leakage

electromagnets. Images obtained with EM are always in shades of gray, and thus EM too deals with the problem of contrast as in the case of optical microscope. In some cases, color can be introduced via computerized software.

There are two types of EMs:

1. Scanning electron microscope (SEM)
2. Transmission electron microscope (TEM)

11.5.1 Scanning electron microscope: Working and biomedical applications

SEM is used to scan the surface of the specimen under observation and produces a 3D image of the surface characteristics of the cells. Before viewing, the specimen has to undergo a relatively long sample preparation step, which involves the following:

a. Fixation of the specimen using aldehyde solution
b. Dehydration of specimen (as it has to be kept in vacuum)
c. Mounting the sample onto a metal stub
d. Coating of the specimen with electrically conductive heavy metal (gold, palladium, or carbon)

Once the sample is prepared, they are kept into a vacuum chamber, and an electron beam is bombarded on it. The electron beam is given off by an electron gun consisting of a tungsten filament that is heated to around 2700 K. The electron beam is focused onto the specimen with the help of condenser lenses that are essentially electromagnets. Voltage used for accelerating electrons ranges between 60 and 100 kV. After being hit with the electron beam, the specimen gives off secondary electrons, Auger electrons, backscattered electrons, X-rays, and photons of various energies. The backscattered electrons and the secondary electrons are used as signals to produce a 3D image of the surface topography of the specimen. When the secondary electrons reach the detector, they strike on a scintillator (a luminescence material that fluoresces when struck by electrons or any high energy particle). This emits rays of light, which get converted into an electric current by a photomultiplier, sending a signal to the cathode ray tube (Fig. 11.4). This produces an image in the similar way a television photograph is generated.

SEM has been resourceful to a wide variety of biomedical research. One of the initial biologic entities to be viewed under SEM were red blood cells. They compared the smooth normal cells with pathologic samples, which could be identified by their granular appearance by using SEM (Salsbury & Clarke, 1967). Following this, there has been a lot of intricate studies on red blood cells (Mustafa et al., 2016; Xu et al., 2012). Examination of articular cartilage surface (Gardner & Woodward, 1968) and

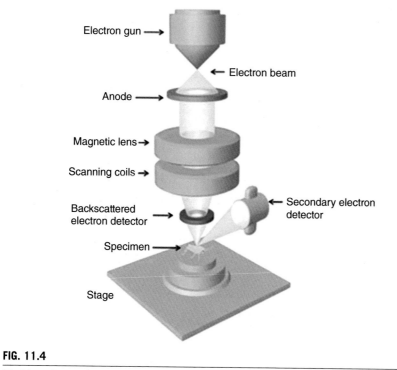

FIG. 11.4

Working of scanning electron microscope.

the changes in fiber patterns and cellular content was studied (McCall, 1968). The morphology and structural properties of dentine have also been revealed by SEM (Boyde & Lester, 1967). There have been various studies of the skin as well. Researchers have presented with normal features of cuticle and cortex of hair and compared them with malformations of the hair (monilethrix and bamboo hair) (Caputo & Ceccarelli, 1969). Different sections of human skin have been studied wherein dermis, epidermis, and the underlying adipose tissue have been examined. The collagen fibers and their alterations in wounds have also been well studied (Forrester et al., 1969). The ultrastructure of surfaces of endothelium and epithelium of cornea have been studied, which reports a clear contrast between the polygonal epithelial cells and the adjacent endothelial cells (Blümcke & Morgenroth, 1967). SEM has also contributed to characterization of micro- and nanostructures, which are slowly becoming major players in health and diagnostic industry. By conjugating proteins, antibodies, or different small biomolecules to nanoparticles, scientists have been able to develop novel drug delivery approaches as well as diagnostic techniques for various diseases (Khanna et al., 2007). It has been used to view the microstructures of various bacterial species (Williams & Davies, 1967). Exosomes obtained from different types of human cells have also been characterized using SEM (Sokolova et al., 2011). SEM are now also available as new generation benchtop versions, which can be used for cell biology and pathology settings (Hyams et al., 2020). Focused ion beam SEM (FIB-SEM) has been quite instrumental in providing good quality images to study the 3D architecture of cells. Recently this technique was combined with ice embedding methodology to ensure that the biologic specimens remain as near to their native state as possible (Spehner et al., 2020). Nanobiotechnology is also gaining loads of useful information on nanostructures using SEM images (Fig. 11.5) (Falsafi et al., 2020; Panwar et al., 2019). SEM has

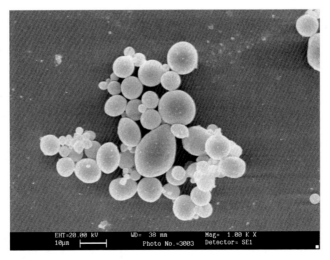

FIG. 11.5

SEM image of polycaprolactone nanoparticles.

proved to be an immensely versatile instrument available for the study and examination of morphology of micro- and nanostructures, and it continues to contribute to the field of biomedical sciences.

11.5.2 Transmission electron microscope: Working and biomedical applications

TEM involves interaction of electron beam with not the surface of the specimen (as in the case of SEM), but it allows for transmission of the electrons through the sample (Fig. 11.6). Therefore, to allow the passage of the electron beam, the samples should be very thin sections. The image is obtained on a fluorescent screen and is 2D, unlike SEM, which produces 3D images. All the other instrumentation and concepts employed in TEM are similar to SEM. TEM offers detailed view of fine structures and internal cellular components.

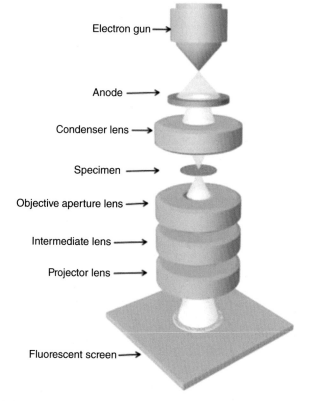

Electron gun

Anode

Condenser lens

Specimen

Objective aperture lens

Intermediate lens

Projector lens

Fluorescent screen

FIG. 11.6

Working of transmission electron microscope.

Preparing the sample for TEM is more extensive than SEM and involves the following steps:

a. Sample fixation using formaldehyde or glutaraldehyde
b. Dehydration of the sample using ethanol
c. Embedding, where ethanol is replaced by an embedding agent, which is a highly miscible plastic
d. Curing at a high temperature (70°C) ensures that the embedding agent polymerizes and becomes solid
e. Thin sections (~90 nm) are then made using glass or diamond knife

Among all the microscopes used for biomedical purposes, TEM is the most powerful with a resolution limit of 0.5 nm. Fig. 11.7 shows a TEM image of a widely used lipid polymer. TEM has contributed and has had a huge impact on research in biomedical sciences. The profound increase in resolution power has allowed viewing of viruses as well. Different families of viruses exhibit morphologic variations, which have facilitated the identification of viruses. TEM has also facilitated discovery of several viruses, notably the viruses linked with gastroenteritis. It is already being employed to examine the new age viruses like SARS and human monkeypox virus. Further, it also plays a vital role in the case of national emergencies as it has the ability to provide frontline diagnostic service as well (Curry et al., 2006). TEM has also been immensely resourceful in characterization studies of nanoparticles and nanoparticle systems. High-resolution TEM imaging is now being combined with atomic resolution electron energy loss spectroscopy, nanodiffraction, and nanometer resolution x-ray energy dispersive spectroscopy techniques. Together these instruments provide critical information that is fundamental to the study of nanoscience and nanotechnology (Liu, 2005). Combination of electron energy loss spectroscopy microanalysis with TEM has been applied to detect calcium during the mineralization

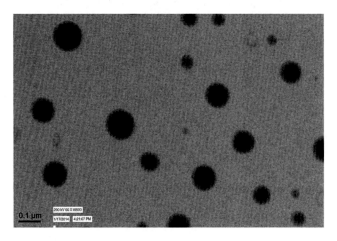

FIG. 11.7

TEM image showing electron dense lipid polymer and electron light surfactant exterior.

process in the bone tissue and also to study the cellular membrane of erythrocytes, which underwent antitumoral drug treatment. This combination has also been used to examine the subcellular localization of an antineoplastic drug in treated samples of human breast tumor cells. This clearly shows that this methodology can be applied for the identification of a target drug (Diociaiuti, 2005). A group studied the crystalline structure and chemical composition of silicate particles that are engulfed in human alveolar macrophages using TEM and found amorphous Si_2AlO_4 and crystalline $ZrSiO_4$ and ferritin (2–4 nm) molecules. TEM has also been used to examine the subcomponents like proteins, lipids, polysaccharides, and nucleic acids of biofilms (Lawrence et al., 2003). Recently introduced techniques like automated scan generators, scanning transmission electron microscopy (STEM), and high-resolution columns have allowed for resolution at subnanometer level. STEM has been used for various pathologic diagnosis like brain, lung, renal, and prostate tissues (Hyams et al., 2020). TEM has also been widely used in the field of nanobiotechnology for studying the morphology of various nanoparticles and nanoformulations (Aidin et al., 2020; Rostamabadi et al., 2020). With the advancements in biomedical research technologies to analyze biomolecules at nanoscale, TEM has been quite instrumental in the whole development process. With the entry of nanotechnology to biomedical sciences in the form of drug delivery tools, molecule tracking, diagnostic tools, among many more, TEM has expanded its usability in the biomedical area more than ever and continues to do so in the coming future.

11.6 Resolution for photon-based and electron-based microscopes

Since the advent of a plethora of microscopes in the market, a fundamental challenge faced by the researchers is how to rate the quality of the microscopes and imaging, so as to decide on which one to use for their studies. Therefore, to understand the limitations of a microscope, the best criteria to be taken under consideration is the resolution of the image. For optical microscopes, resolution is described by Abbe equation.

$$d = \frac{0.612 * 1}{n \sin a}$$

where:

> d = resolution
> l = wavelength of illuminating radiation
> n = index of refraction of medium between point source and lens
> a = half the angle of the cone of light coming from specimen plane that is accepted by the objective (in radians)
> $n \sin a$ is expressed as numeric aperture

In Abbe equation, d refers to the minimum distance between two objects such that they can be distinguished as single objects. It depends on the wavelength of the light being illuminated on the sample and the numerical aperture, which in turn depends on the ability of the objective lens to gather light. A lens usually works in air ($n = 1.00$) or oil ($n = 1.25$). Oil allows collection of a larger portion of light. Abbe equation gives the diffraction-limited resolution of a microscope. If all the distortions and aberrations are removed from the system, d will give the correct value of resolution.

To describe the resolution of an EM, the illuminating wavelength has to be first described in terms of voltage, which is represented as:

$$l = \frac{1.23\text{nm}}{V^{1/2}}$$

where,

l = wavelength
V = potential difference under which electrons are passing

On substituting the value of l (wavelength) to Abbe equation, we get:

$$d = \frac{0.753}{aV^{1/2}}$$

where:

d = resolution in nm
a = half aperture angle
V = potential difference

Now, if we solve for 100,000 volts, we get a theoretical value of 0.24 nm. This value improves with higher accelerating voltages. As with the case of optical microscopes, each aperture and lens in EMs too have distortions and aberrations. Thus the practical limit of resolution shows a higher value than the one obtained theoretically.

Both the photon-based microscopes and electron-based microscopes have come up with high-quality imaging that has offered better insights into the world of biomedical sciences. With time, the development of various sophisticated softwares have added to manipulating the images into 3D, which has helped in analysis to various internal cellular structures. Advancements in microscopy will widen the knowledge of morphology, topography, and structural information of cell organelles. Thus, it is an important field for the development of biomedical sciences.

11.7 Conclusion

Researchers have come a long way from mere observation of stained specimens under a microscope with the advent of better technologies in not only microscope hardware but computational studies as well as chemical engineering. Research groups can now not only visualize stationary samples but can perform live tracking of molecules, which has proved to be an immensely useful tool in the field of cell biology and pathology. Advancements in TEM have allowed for resolution as low as 0.5 nm, which has facilitated observations of nanostructures, which was impossible to do just a few years back. New age fluorescence and confocal microscopy techniques have facilitated study of protein localization, function, and interaction in living cells. Table 11.3 gives an overall summary of the specifications of each of the microscopes mentioned in this chapter. When appropriately used, each type of microscope has proved to be a valuable tool for both quantitative and qualitative analysis of biologic samples. Moreover, easy availability of these specialized microscopy techniques has broadened the associated research community and will continue to add value to the plethora of information already available on biomedical sciences.

Table 11.3 Features of different types of microscopy.

Microscope	Illumination Source	Resolution	Magnification	Specimen	Features
Bright-field Microscope	Light	200 nm	1000X	Live and stained specimens	Dark specimen against a bright background Provides fair morphology information
Dark-field Microscope	Light	200 nm	1000X	Live and unstained specimens	Bright specimen against a dark background Provides the outline of the cellular structure without sharper image of cellular components
Phase Contrast Microscope	Light	200 nm	1000X	Live specimens	Image of specimen is obtained against a gray background Gives good details of internal cellular components
Differential Interference Microscope	Light	200 nm	1000X	Live specimens	Gives 3D, colored, and highly contrasting image of specimens
Fluorescence Microscope	Light/Ultraviolet light	100–200 nm	1000X	Specimens stained with fluorescent dyes or combined with fluorescent antibodies	Provides colored and contrasting image Good to differentiate between several cellular components Excellent diagnostic tool

Microscope	Illumination Source	Reso-lution	Magnifica-tion	Specimen	Features
Confocal Microscope	Light/Laser	10 nm	1500X	Specimens stained with fluores-cent dyes or com-bined with fluorescent antibodies	High specificity and 3D images with high contrast
Near-field Scanning Optical Mi-croscope	Light	20 nm	2000X	Live and fluorescently stained specimens	Aperture scan-ning of line grafting
Transmis-sion Electron Microscope	Electron beam	0.5 nm	1,000,000X	Dead or dried specimens coated with heavy metals	Finest details of internal structures of cell and viruses
Scanning Electron Microscope	Electron beam	2–5 nm	1,000,000X	Dead or dried specimens coated with heavy metals	Detailed external structure and cellular arrange-ments

References

Aidin, M., Tafreshi, H., Ali, H., 2020. The study of biofunctionalization of carbon nanotubes and their applications in biology. Medbiotech Journal 4 (2), 70–82.

Altun, A.A., Taghiyev, A., 2017. Advanced image processing techniques and applications for biological objects, 2017 2nd IEEE International Conference on Computational Intelligence and Applications (ICCIA). IEEE, pp. 340–344.

Ash, E.A., Nicholls, G., 1972. Super-resolution aperture scanning microscope. Nature 237 (5357), 510–512.

Axelrod, D., Koppel, D.E., Schlessinger, J., Elson, E., Webb, W.W., 1976. Mobility measurement by analysis of fluorescence photobleaching recovery kinetics. Biophysical Journal 16 (9), 1055.

Bai, Z., Liu, Y., Kong, R., Nie, T., Sun, Y., Li, H., … Song, Q., 2020. Near-field Terahertz sensing of HeLa cells and pseudomonas based on monolithic integrated metamaterials with a spintronic Terahertz Emitter. ACS Applied Materials & Interfaces 12 (32), 35895–35902.

Batchelor, H., Dettmar, P., Hampson, F., Jolliffe, I., Craig, D., 2004. Microscopic techniques as potential tools to quantify the extent of bioadhesion of liquid systems. European Journal of Pharmaceutical Sciences 22 (5), 341–346.

Blümcke, S., Morgenroth Jr, K., 1967. The stereo ultrastructure of the external and internal surface of the cornea. Journal of Ultrastructure Research 18 (5-6), 502–518.

Boyde, A., 1985. Stereoscopic images in confocal (tandem scanning) microscopy. Science 230 (4731), 1270–1272.

Boyde, A., Lester, K.S., 1967. Electron microscopy of resorbing surfaces of dental hard tissues. Zeitschrift Ffür Zellforschung und Mikroskopische Anatomie 83 (4), 538–548.

Brissova, M., Fowler, M.J., Nicholson, W.E., Chu, A., Hirshberg, B., Harlan, D.M., Powers, A.C., 2005. Assessment of human pancreatic islet architecture and composition by laser scanning confocal microscopy. Journal of Histochemistry & Cytochemistry 53 (9), 1087–1097.

Caputo, R., Ceccarelli, B., 1969. Study of normal hair and of some malformations with a scanning electron microscope. Archiv Für Klinische und Experimentelle Dermatologie 234 (3), 242–249.

Caspers, P.J., Lucassen, G.W., Puppels, G.J., 2003. Combined in vivo confocal Raman spectroscopy and confocal microscopy of human skin. Biophysical Journal 85 (1), 572–580.

Chen, H., Sedat, J.W., Agard, D.A., 1990. Manipulation, display, and analysis of three-dimensional biological imagesHandbook of biological confocal microscopy. Springer, pp. 141–150.

Curry, A., Appleton, H., Dowsett, B., 2006. Application of transmission electron microscopy to the clinical study of viral and bacterial infections: present and future. Micron 37 (2), 91–106.

Davidovits, P., Egger, M.D., 1969. Scanning laser microscope. Nature 223 (5208), 831.

de Paula, A.M., Toledo, J.A., Silva, H.B., Weber, G., 2001. Near-field scanning optical images of bacteria, MRS Online Proceedings Library Archive, 711.

Diociaiuti, M., 2005. Electron energy loss spectroscopy microanalysis and imaging in the transmission electron microscope: example of biological applications. Journal of Electron Spectroscopy and Related Phenomena 143 (2-3), 189–203.

Dundr, M., Hoffmann-Rohrer, U., Hu, Q., Grummt, I., Rothblum, L.I., Phair, R.D., Misteli, T., 2002. A kinetic framework for a mammalian RNA polymerase in vivo. Science 298 (5598), 1623–1626.

Dunn, G.A., Dobbie, I.M., Monypenny, J., Holt, M.R., Zicha, D., 2002. Fluorescence localization after photobleaching (FLAP): A new method for studying protein dynamics in living cells. Journal of Microscopy 205 (1), 109–112.

Enderle, T., Ha, T., Chemla, D.S., Weiss, S., 1998. Near-field fluorescence microscopy of cells. Ultramicroscopy 71 (1-4), 303–309.

Falsafi, S.R., Rostamabadi, H., Assadpour, E., Jafari, S.M., 2020. Morphology and microstructural analysis of bioactive-loaded micro/nanocarriers via microscopy techniques; CLSM/SEM/TEM/AFM. Advances in Colloid and Interface Science, 102166.

Forrester, J.C., Hunt, T.K., Hayes, T.L., Pease, R.F.W., 1969. Scanning electron microscopy of healing wounds. Nature 221 (5178), 373–374.

Gambhir, S.S., Bauer, E., Black, M.E., Liang, Q., Kokoris, M.S., Barrio, J.R., … Herschman, H.R., 2000. A mutant herpes simplex virus type 1 thymidine kinase reporter gene shows improved sensitivity for imaging reporter gene expression with positron emission tomography, Proceedings of the National Academy of Sciences, 97, 2785–2790.

Garcia-Parajo, M.F., Veerman, J.A., Ruiter, A.G.T., Van Hulst, N.F., 1998. Near-field optical and shear-force microscopy of single fluorophores and DNA molecules. Ultramicroscopy 71 (1-4), 311–319.

Gardner, D.L., Woodward, D.H., 1968. Scanning electron microscopy of articular surfaces. The Lancet 292 (7580), 1246.

Geng, G., Dai, G., Li, D., Zhou, S., Li, Z., Yang, Z., … Wang, H., 2019. Imaging brain tissue slices with terahertz near-field microscopy. Biotechnology Progress 35 (2), e2741.

Gratton, E., VandeVen, M.J., 1989. Laser sources for confocal microscopy. In: Pawley, J.B. (Ed.), Handbook of biological confocal microscopy. Plenum Press, pp. 53–67.

Ha, T., Enderle, T., Chemla, D.S., Weiss, S., 1996. Dual-molecule spectroscopy: molecular rulers for the study of biological macromolecules. IEEE Journal of Selected Topics in Quantum Electronics 2 (4), 1115–1128.

Huzaira, M., Rius, F., Rajadhyaksha, M., Anderson, R.R., González, S., 2001. Topographic variations in normal skin, as viewed by in vivo reflectance confocal microscopy. Journal of Investigative Dermatology 116 (6), 846–852.

Hyams, T.C., Mam, K., Killingsworth, M.C., 2020. Scanning electron microscopy as a new tool for diagnostic pathology and cell biology. Micron 130, 102797.

Inoue, H., Igari, T., Nishikage, T., Ami, K., Yoshida, T., Iwai, T., 2000. A novel method of virtual histopathology using laser-scanning confocal microscopy in-vitro with untreated fresh specimens from the gastrointestinal mucosa. Endoscopy 32 (06), 439–443.

Jacobson, K., Ishihara, A., Inman, R., 1987. Lateral diffusion of proteins in membranes. Annual Review of Physiology 49 (1), 163–175.

Jares-Erijman, E.A., Jovin, T.M., 2003. FRET imaging. Nature Biotechnology 21 (11), 1387–1395.

Kandel, M.E., He, Y.R., Lee, Y.J., Chen, T.H.Y., Sullivan, K.M., Aydin, O., … Popescu, G., 2020. Phase imaging with computational specificity (PICS) for measuring dry mass changes in sub-cellular compartments. Nature Communications 11 (1), 1–10.

Keller, T.H., Rayment, T., Klenerman, D., 1998. Optical chemical imaging of tobacco mosaic virus in solution at 60-nm resolution. Biophysical Journal 74 (4), 2076–2079.

Khanna, P.K., Gaikwad, S., Adhyapak, P.V., Singh, N., Marimuthu, R., 2007. Synthesis and characterization of copper nanoparticles. Materials Letters 61 (25), 4711–4714.

Kimura, H., Hieda, M., Cook, P.R., 2003. Measuring histone and polymerase dynamics in living cellsMethods in enzymology375. Academic Press, pp. 381–393.

Koppel, D.E., Axelrod, D., Schlessinger, J., Elson, E.L., Webb, W.W., 1976. Dynamics of fluorescence marker concentration as a probe of mobility. Biophysical Journal 16 (11), 1315–1329.

Kvaal, S., Solheim, T., 1989. Fluorescence from dentin and cementum in human mandibular second premolars and its relation to age. European Journal of Oral Sciences 97 (2), 131–138.

Lawrence, J.R., Swerhone, G.D.W., Leppard, G.G., Araki, T., Zhang, X., West, M.M., Hitchcock, A.P., 2003. Scanning transmission X-ray, laser scanning, and transmission electron microscopy mapping of the exopolymeric matrix of microbial biofilms. Applied and Environmental Microbiology 69 (9), 5543–5554.

Liebman, P.A., Entine, G., 1974. Lateral diffusion of visual pigment in photoreceptor disk membranes. Science 185 (4149), 457–459.

Liu, J., 2005. Scanning transmission electron microscopy and its application to the study of nanoparticles and nanoparticle systems. Journal of Electron Microscopy 54 (3), 251–278.

Livet, J., Weissman, T.A., Kang, H., Draft, R.W., Lu, J., Bennis, R.A., … Lichtman, J.W., 2007. Transgenic strategies for combinatorial expression of fluorescent proteins in the nervous system. Nature 450 (7166), 56–62.

McCall, J.G., 1968. Scanning electron microscopy of articular surfaces. Lancet (London, England) 2 (7579), 1194.

Miller, D.S., Nobmann, S.N., Gutmann, H., Toeroek, M., Drewe, J., Fricker, G., 2000. Xenobiotic transport across isolated brain microvessels studied by confocal microscopy. Molecular Pharmacology 58 (6), 1357–1367.

Miyawaki, A., Llopis, J., Heim, R., McCaffery, J.M., Adams, J.A., Ikura, M., Tsien, R.Y., 1997. Fluorescent indicators for Ca 2+ based on green fluorescent proteins and calmodulin. Nature 388 (6645), 882–887.

Moers, M.H.P., Kalle, W.H.J., Ruiter, A.G.T., Wiegant, J.C.A.G., Raap, A.K., Greve, J., ... Van Hulst, N.F., 1996. Fluorescence in situ hybridization on human metaphase chromosomes detected by near-field scanning optical microscopy. Journal of Microscopy 182 (1), 40–45.

Monetta, P., Slavin, I., Romero, N., Alvarez, C., 2007. Rab1b interacts with GBF1 and modulates both ARF1 dynamics and COPI association. Molecular Biology of the Cell 18 (7), 2400–2410.

Muramatsu, H., Chiba, N., Ataka, T., Iwabuchi, S., Nagatani, N., Tamiya, E., Fujihira, M., 1996. Scanning near-field optical/atomic force microscopy for fluorescence imaging and spectroscopy of biomaterials in air and liquid: observation of recombinant Escherichia coli with gene coding to green fluorescent protein. Optical Review 3 (6B), 470–474.

Mustafa, I., Al Marwani, A., Mamdouh Nasr, K., Abdulla Kano, N., Hadwan, T., 2016. Time dependent assessment of morphological changes: leukodepleted packed red blood cells stored in SAGM. BioMed Research International *2016*.

Nehls, S., Snapp, E.L., Cole, N.B., Zaal, K.J., Kenworthy, A.K., Roberts, T.H., ... Lippincott-Schwartz, J., 2000. Dynamics and retention of misfolded proteins in native ER membranes. Nature Cell Biology 2 (5), 288–295.

Panwar, N., Soehartono, A.M., Chan, K.K., Zeng, S., Xu, G., Qu, J., ... Chen, X., 2019. Nanocarbons for biology and medicine: Sensing, imaging, and drug delivery. Chemical Reviews 119 (16), 9559–9656.

Peters, R., Beck, K., 1983. Translational diffusion in phospholipid monolayers measured by fluorescence microphotolysis. Proceedings of the National Academy of Sciences 80 (23), 7183–7187.

Petroll, W.M., Cavanagh, H.D., Lemp, M.A., Andrews, P.M., Jester, J.V., 1992. Digital image acquisition in in vivo confocal microscopy. Journal of Microscopy 165 (1), 61–69.

Petty, H.R., 2007. Fluorescence microscopy: established and emerging methods, experimental strategies, and applications in immunology. Microscopy Research and Technique 70 (8), 687–709.

Phair, R.D., Misteli, T., 2000. High mobility of proteins in the mammalian cell nucleus. Nature 404 (6778), 604–609.

Pygall, S.R., Whetstone, J., Timmins, P., Melia, C.D., 2007. Pharmaceutical applications of confocal laser scanning microscopy: The physical characterisation of pharmaceutical systems. Advanced Drug Delivery Reviews 59 (14), 1434–1452.

Reits, E.A., Neefjes, J.J. (2001). From fixed to FRAP: Measuring protein mobility and activity in living cells. Nature Cell Biology 3 (6), E145.

Rostamabadi, H., Falsafi, S.R., Jafari, S.M., 2020. Transmission electron microscopy (TEM) of nanoencapsulated food ingredients. Characterization of nanoencapsulated food ingredients. Academic Press, pp. 53–82.

Roysam, B., Bhattacharjya, A.K., Srinivas, C., Szarowski, D.H., Turner, J.N., 1992. Unsupervised noise removal algorithms for 3-D confocal fluorescence microscopyBiomedical image processing and three-dimensional microscopy1660. International Society for Optics and Photonics, pp. 250–261.

Salsbury, A.J., Clarke, J.A., 1967. New method for detecting changes in the surface appearance of human red blood cells. Journal of Clinical Pathology 20 (4), 603–610.

Selkin, B., Rajadhyaksha, M., Gonzalez, S., Langley, R.G., 2001. In vivo confocal microscopy in dermatology. Dermatologic Clinics 19 (2), 369–377.

Shapiro, M.G., Goodwill, P.W., Neogy, A., Yin, M., Foster, F.S., Schaffer, D.V., Conolly, S.M., 2014. Biogenic gas nanostructures as ultrasonic molecular reporters. Nature Nanotechnology 9 (4), 311.

Sliney, D., Wolbarsht, M., 1980. Current laser exposure limits, safety with lasers and other optical sources, a comprehensive handbook. Plenum Press, pp. 261–283.

Smith, S., Monson, E.E., Merritt, G., Tan, W., Birnbaum, D., Shi, Z.Y., … Merlin, R.D., 1993. Tip/sample interactions, contrast, and near-field microscopy of biological and solid-state samplesScanning probe microscopies II1855. International Society for Optics and Photonics, pp. 81–92.

Sokolova, V., Ludwig, A.K., Hornung, S., Rotan, O., Horn, P.A., Epple, M., Giebel, B., 2011. Characterisation of exosomes derived from human cells by nanoparticle tracking analysis and scanning electron microscopy. Colloids and Surfaces B: Biointerfaces 87 (1), 146–150.

Spehner, D., Steyer, A.M., Bertinetti, L., Orlov, I., Benoit, L., Pernet-Gallay, K., … Schultz, P., 2020. Cryo-FIB-SEM as a promising tool for localizing proteins in 3D. Journal of Structural Biology, 107528.

Subramaniam, V., Kirsch, A.K., Rivera-Pomar, R.V., Jovin, T.M., 1997. Scanning near-field optical microscopy and microspectroscopy of green fluorescent protein in intact Escherichia coli bacteria. Journal of Fluorescence 7 (4), 381–385.

Sundström, F., Fredriksson, K., Montan, S., Hafström-Björkman, U., Ström, J., 1985. Laser-induced fluorescence from sound and carious tooth substance: spectroscopic studies. Swedish Dental Journal 9 (2), 71–80.

Synge, E., 1928. XXXVIII. A suggested method for extending microscopic resolution into the ultra-microscopic region. The London, Edinburgh, and Dublin Philosophical Magazine and Journal of Science 6 (35), 356–362.

Talley, C.E., Cooksey, G.A., Dunn, R.C., 1996. High resolution fluorescence imaging with cantilevered near-field fiber optic probes. Applied Physics Letters 69 (25), 3809–3811.

Tamiya, E., Iwabuchi, S., Nagatani, N., Murakami, Y., Sakaguchi, T., Yokoyama, K., … Muramatsu, H., 1997. Simultaneous topographic and fluorescence imagings of recombinant bacterial cells containing a green fluorescent protein gene detected by a scanning near-field optical/atomic force microscope. Analytical Chemistry 69 (18), 3697–3701.

Tamm, L.K., Böhm, C., Yang, J., Shao, Z., Hwang, J., Edidin, M., Betzig, E., 1996. Nanostructure of supported phospholipid monolayers and bilayers by scanning probe microscopy. Thin Solid Films 284, 813–816.

Tan, C., Guan, Y., Feng, Z., Ni, H., Zhang, Z., Wang, Z., … Li, A., 2020. Deepbrainseg: Automated brain region segmentation for micro-optical images with a convolutional neural network. Frontiers in Neuroscience, 14.

Thong, P.S.P., Olivo, M.C., Kho, K.W., Mancer, K., Zheng, W., Harris, M.R., Soo, K.C., 2007. Laser confocal endomicroscopy as a novel technique for fluorescence diagnostic imaging of the oral cavity. Journal of Biomedical Optics 12 (1), 014007.

Tocanne, J.F., Dupou-Cézanne, L., Lopez, A., Tournier, J.F., 1989. Lipid lateral diffusion and membrane organization. FEBS Letters 257 (1), 10–16.

Turner, J.N., Ancin, H., Becker, D.E., Szarowski, D.H., Holmes, M., O'Connor, N., … Roysam, B., 1997. Automated image analysis technologies for biological 3D light microscopy. International Journal of Imaging Systems and Technology 8 (3), 240–254.

Umakoshi, T., Fukuda, S., Iino, R., Uchihashi, T., Ando, T., 2020. High-speed near-field fluorescence microscopy combined with high-speed atomic force microscopy for biological studies. Biochimica et Biophysica Acta (BBA)-General Subjects *1864* (2), 129325.

Watson, T.F., 1997. Fact and artefact in confocal microscopy. Advances in Dental Research 11 (4), 433–441.

Weissleder, R., Moore, A., Mahmood, U., Bhorade, R., Benveniste, H., Chiocca, E.A., Basilion, J.P., 2000. In vivo magnetic resonance imaging of transgene expression. Nature Medicine 6 (3), 351–354.

White, J., Stelzer, E., 1999. Photobleaching GFP reveals protein dynamics inside live cells. Trends in Cell Biology 9 (2), 61–65.

White, N.S., Errington, R.J., 2005. Fluorescence techniques for drug delivery research: theory and practice. Advanced Drug Delivery Reviews 57 (1), 17–42.

Williams, S.T., Davies, F.L., 1967. Use of a scanning electron microscope for the examination of actinomycetes. Microbiology 48 (2), 171–177.

Wilson, T., 1989. Three-dimensional imaging in confocal systems. Journal of Microscopy 153, 161–169.

Winburn, D.C., 1985. Laser-eye damage thresholds, occupational safety and health: Practical laser safety. Marcel Dekker, pp. 19–26.

Wright, S.J., Schatten, G., 1991. Confocal fluorescence microscopy and three-dimensional reconstruction. Journal of Electron Microscopy Technique 18 (1), 2–10.

Wright, S.J., Centonze, V.E., Stricker, S.A., DeVries, P.J., Paddock, S.W., Schatten, G., 1993. Introduction to confocal microscopy and three-dimensional reconstructionMethods in cell biology38. Academic Press, pp. 1–45.

Xu, D., Peng, M., Zhang, Z., Dong, G., Zhang, Y., Yu, H., 2012. Study of damage to red blood cells exposed to different doses of γ-ray irradiation. Blood Transfusion 10 (3), 321.

Index

Page numbers followed by "*f*" and "*t*" indicate, figures and tables respectively.